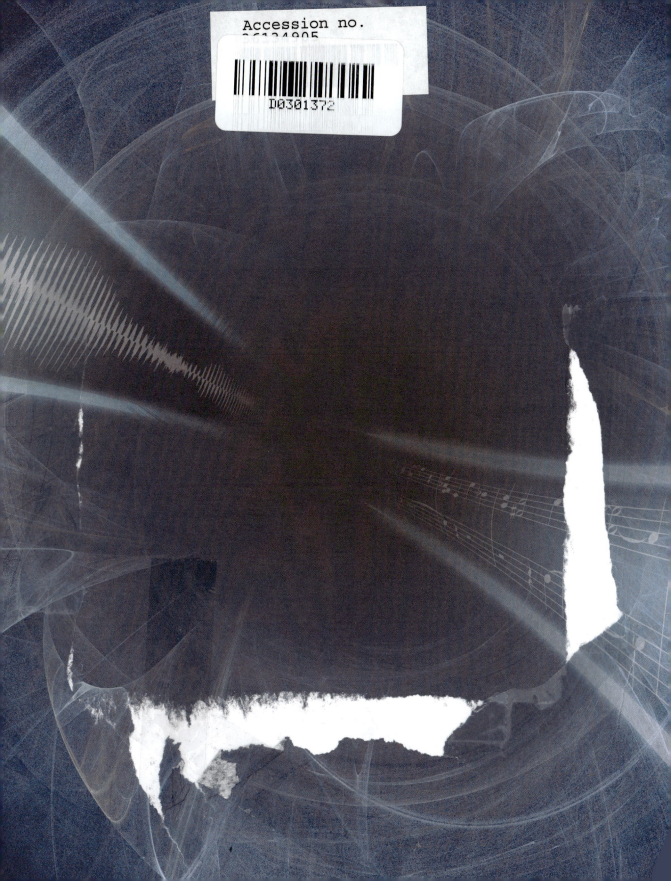

AUDIO PRODUCTION AND CRITICAL LISTENING
Technical Ear Training

JASON COREY

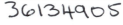

ELSEVIER

AMSTERDAM • BOSTON • HEIDELBERG • LONDON • NEW YORK • OXFORD
PARIS • SAN DIEGO • SAN FRANCISCO • SINGAPORE • SYDNEY • TOKYO
Focal Press is an imprint of Elsevier

Focal Press

Focal Press is an imprint of Elsevier
30 Corporate Drive, Suite 400, Burlington, MA 01803, USA
The Boulevard, Langford Lane, Kidlington, Oxford, OX5 1GB, UK

Library of Congress Cataloging-in-Publication Data
Corey, Jason.
Audio production and critical listening : technical ear training/Jason Corey.
    p. cm.
Includes bibliographical references and index.
ISBN 978-0-240-81295-3 (alk. paper)
1. Auditory perception. 2. Listening. 3. Ear training. 4. Sound—Recording and reproducing—Evaluation. I. Title.
TA365.C678 2010
781.49—dc22

        2009052978

British Library Cataloguing-in-Publication Data
A catalogue record for this book is available from the British Library.

ISBN: 978-0-240-81295-3

10 11   5 4 3 2 1

Printed in China

CONTENTS

Acknowledgements

This book would not have come about without the direct and indirect assistance of many people.

First, I want to thank all those who have imparted their wisdom and experience about critical listening skills to me, most notably: Peter Cook, John Klepko, Geoff Martin, Tim Martyn, George Massenburg, René Quesnel, and Wieslaw Woszczyk.

Thank you to those who helped with the preparation of the manuscript: Steve Bellamy, Justin Crowell, and Tim Sullivan.

Thank you to Christopher Kendall, Mary Simoni, Rebecca Sestili, and the Office of the Vice-President for Research at the University of Michigan, as well as Theresa Leonard and The Banff Centre for invaluable support at various points throughout this project.

Thank you to Mike Halerz for the cover design.

Thank you to students at the University of Michigan and McGill University who provided feedback and suggestions during the development of the ear training software modules.

Thank you to Catharine Steers, Carlin Reagan, Tahya Pawnell, and Laura Aberle at Focal Press for editorial guidance and software testing.

Finally, I would like to thank my wonderful wife, Jennifer, for her love and support.

INTRODUCTION

The practice of audio engineering is both an art and a science. For success in audio production, an engineer should ideally possess both an understanding of theoretical concepts and highly developed critical listening skills related to sound recording and production. Each recording project has its own set of requirements, and engineers cannot rely on one set of recording procedures for every project. As such, they must rely on a combination of technical knowledge and listening skills to guide their work.

Although technical knowledge of analog electronics, digital signal processing, audio signal analysis, and the theoretical aspects of audio equipment is critical for a solid understanding of the principles of audio engineering, many of the decisions made during a recording project—such as microphone choice and location, mix balance, fader levels, and signal processing—are based solely on what is heard. As such, it is often the ability to navigate the subjective impressions of audio that allows engineers to successfully improve on sound quality.

Every action taken by an engineer in relation to an audio signal will have some effect on the sound that a listener hears, however subtle it may be, and an engineer must have an attentive ear tuned to the finest details of timbre and sound quality. Most of these subjective decisions are in response to the artistic goals of a project, and engineers must determine, based on what they hear, if a technical choice is contributing to or detracting from these goals. Engineers need to know how the technical parameters of audio hardware and software devices affect perceived sonic attributes.

In addition to possessing technical and theoretical expertise, successful audio engineers possess the capacity to differentiate timbral, dynamic, and technical details of sound. They can translate their aural impressions into appropriate technical judgments and alterations. Sometimes referred to as "Golden Ears," these highly experienced audio professionals possess the extraordinary ability to focus their auditory attention, resulting in the efficient and accurate control of audio signals. They are expert listeners, individuals who possess highly developed critical listening skills and who can identify fine details of sound and make consistent judgments about what they hear (Stone, 1993). Such experienced engineers identify shortcomings to be solved and features to be highlighted in an audio signal.

Critical listening skills in audio can be developed and improved gradually over time by engineers as they work in the field of audio, but there are systematic methods that can shorten the time required to make significant progress in ear training. As René Quesnel reported in his doctoral thesis, sound recording students who completed systematic technical ear training outperformed experienced audio professionals on tasks such as identifying frequency and gain settings of parametric equalization (Quesnel, 2001). Typically, the development of listening skills for an audio engineer happens on the job. Although it was once common for beginning engineers to work with more experienced engineers and learn from them in the context of practical experience, the audio industry has gone through drastic changes and the apprentice model is gradually disappearing from the practice of audio engineering. Despite this evolution in the audio industry, critical listening skills remain as important as ever, especially as we see audio quality decline in many consumer audio formats. This book presents some ideas for developing critical listening skills and potentially reducing the time it takes to develop them.

A number of questions emerge as we begin to consider critical listening skills related to sound recording and production:

- What aural skills do experienced sound engineers, producers, tonmeisters, and musicians possess that allow them to make recordings, mix sound for film, or equalize sound systems better than a novice engineer?
- What can the legendary engineers and producers, who have extraordinary abilities to identify and manipulate sonic timbres, hear that the average person cannot?
- How do audio professionals hear and consistently identify extremely subtle features or changes in an audio signal?

- How do expert listeners translate between their perceptions of sound and the physical control parameters available to them?
- How can nonexpert listeners gain similar skills, allowing them to identify the physical parameters of an audio signal necessary to achieve a desired perceptual effect?
- What specific aspects of sound should novice audio engineers be listening for?

There has been a significant amount written on the technical and theoretical aspects of sound, sound reproduction, and auditory perception, but this book focuses on the development of the critical listening skills necessary for the successful practice of audio engineering.

To facilitate the training process, software modules accompanying the book allow the reader to practice listening to the effects of different types of audio signal processing. The software practice modules allow a progression through various levels of difficulty and provide necessary practical training in the development of technical listening skills.

Audio Attributes

The main objective of this book and accompanying software is to explore critical listening as it relates to typical types of audio signal processing. Distinct from musical aural skills or solfège, technical ear training focuses on the sonic effects of the most common types of signal processing used in sound recording and sound reproduction systems, such as equalization, dynamics processing, and reverberation. Knowledge of the sonic effects of audio signal processing, coupled with the ability to discriminate between small changes in sound quality, allow engineers to make effective changes to reproduced sound as needed. Highly developed critical listening skills allow an engineer to identify not only the effects of deliberate signal processing but also the unintentional or unwanted artifacts such as noise, buzz, hum, and distortion. Once such undesirable sounds are identified, an engineer can work to eliminate or reduce their presence.

The book is organized according to common audio processing tools available to the audio engineer. In this book we will explore the following main audio attributes and associated devices:

- Spectral balance—parametric equalization
- Spatial attributes—delay and reverberation
- Dynamic range control—compression/limiting and expansion
- Sounds or qualities of sound that can detract from recordings—distortion and noise
- Audio excerpt cut-off points—source-destination editing

Goals of the Book

There are three main goals of this book and software:

1. To facilitate isomorphic mapping of technical parameters and perceived qualities of sound. Isomorphic mapping is a linking of technical and engineering parameters to auditory perceptual attributes. Engineers need to be able to diagnose problematic sonic artifacts in a recording and have an understanding of their causes. In audio, engineers are translating between physical control parameters (i.e., frequency in Hertz, sound level in decibels) and the perception of an audio signal (i.e., timbre, loudness).
2. To heighten awareness of subtle features and attributes of sound, and to promote a greater ability to differentiate among minute changes in sound quality or signal processing.
3. To increase the speed with which one can identify features of sound, translate between auditory perceptions and signal processing control parameters, and decide on what physical parameters need to be changed in a given situation.

To achieve these goals Chapters 2, 3, 4, and 5 focus on specific types of audio processing and artifacts: equalization, reverberation and delay, dynamics processing, and distortion and noise, respectively.

Chapter 2 focuses on the spectral balance of an audio signal and how that is influenced by filtering and parametric equalization. The spectral balance is the relative level of various frequency bands within the full audio band (from 20 to 20,000 Hz), and this chapter focuses specifically on parametric equalizers.

The spatial properties of reproduced sound include the panning of sources, reverberation, echo, and delay (with and without feedback). Chapter 3 examines training methods for spatial attributes.

Dynamics processing is used widely in recorded music. Audio processing effects such as compression, limiting, expansion, and gating all offer means to sculpt audio signals in unique and time-varying ways. Dynamic range compression may be one of the most difficult types of processing for a beginning engineer to learn how to use. On many algorithms, the controllable parameters are interrelated to a certain extent and affect how they are used and heard. Chapter 4 takes a look at dynamic processing and offers practice exercises on hearing artifacts produced by these different effects.

Distortion can be applied intentionally to a recording or elements within a recording as an effect such as with electric guitars, but recording engineers generally try to avoid unintentional distortion such as through overloading an analog gain stage or analog-to-digital converter. Chapter 5 explores additional types of distortion, such as bit rate reduction and perceptual encoding, as well as other types of sonic artifacts that detract from a sound recording, namely extraneous noises, clicks, pops, buzz, and hum.

Chapter 6 focuses on audio excerpt cut-off points and introduces a novel type of ear-training practice based on the process of source-destination editing. The act of finding edit points can also sharpen the ability to differentiate changes in cut-off points at the millisecond level. The accompanying software module mimics the process of finding an edit point by comparing the end point of one clip with the end point of a second clip of identical music.

Finally, Chapter 7 examines analysis techniques for recorded sound. Although there are established traditions of the theoretical analysis of music, there is no standardized method of analyzing recordings from a timbral, sound quality, spatial image, aesthetic, or technical point of view. This chapter presents some methods for analyzing musical recordings and presents some examples of analysis of commercially available recordings.

There have been significant contributions to the field of technical ear training appearing in conference and journal articles, including Bech's "Selection and Training of Subjects for Listening Tests on Sound-Reproducing Equipment" (1992); Kassier, Brookes, and Rumsey's "Training versus Practice in Spatial Audio Attribute Evaluation Tasks" (2007); Miskiewicz's "Timbre Solfege: A Course in Technical Listening for Sound Engineers" (1992); Olive's "A Method for Training Listeners and Selecting Program Material for Listening Tests" (1994); and Quesnel's "Timbral Ear-Trainer: Adaptive, Interactive Training of Listening Skills for Evaluation of Timbre" (1996). This book draws from previous research and presents methods for practice and development of critical listening skills in the context of audio production.

The author assumes that the reader has completed some undergraduate-level study in sound recording theory and practice and has an understanding of basic audio theory topics such as decibels, equalization, dynamics, microphones, and microphone techniques.

The Accompanying Software

Because of the somewhat abstract nature of simply reading about critical listening, a number of software modules have been included with this book to help the reader practice hearing various types of signal processing that are described herein. The accompanying software practice modules are interactive, allowing the user to adjust parameters of each type of processing and be given immediate auditory feedback, mimicking what happens in the recording and mixing studio. Although some of the modules simply provide sound processing examples, others offer exercises involving matching and absolute identification of processing parameters by ear. The benefit of matching exercises lies mostly in providing the opportunity to rely completely on what is heard without having to translate to a verbal representation of a sound.

The use of digital recordings for ear training practice has an advantage over analog recordings or acoustical sounds in that digital recordings can be played back numerous times in exactly the same way. Some specific sound recordings are suggested in the book, but there are other locations to obtain useful sound samples for focusing on different types of processing. As of this writing, single instrument samples and stem mixes can be downloaded from many web sites, such as the following:

http://bush-of-ghosts.com/remix/bush_of_ghosts.htm
www.freesound.org
www.realworldremixed.com/download.php
www.royerlabs.com

Furthermore, software programs such as Apple's Logic and GarageBand include libraries of single instrument sounds that can serve as sound sources in the software practice modules.

This book does not focus on specific models of commercially available audio processing software or hardware but treats each type of processing as typical of what may be found among professional audio devices and software. Audio processing modules that are available commercially vary from one model to the next, and the author feels that the training discussed in this book and applied in the software modules serve as a solid starting point for ear training and can be extrapolated to most commercial models.

What this book does not attempt to do is provide recommendations for signal processing settings or microphone techniques for different instruments or recording setups. It is impossible to have a one-size-fits-all approach to audio production, and the goal is help the reader to listen more critically and with more detail in order to shape each individual recording.

All of the software modules are included on the accompanying CD-ROM, and updates to the software will be posted periodically on the author's web page: www-personal.umich.edu/~coreyja.

LISTENING

CHAPTER OUTLINE

Audio Production and Critical Listening. DOI: 10.1016/B978-0-240-81295-3.00001-0

We are exposed to sound throughout each moment of every day regardless of whether we pay attention to it or not. The sounds we hear give insight into not only their sources but also the nature of our physical environment—surrounding objects, walls, and structures. Whether we find ourselves in a highly reverberant environment or an anechoic chamber, the quality of reflected sound or the lack of reflections informs us about the physical properties of our location. Our surrounding environment becomes audible, even if it is not creating sound itself, by the way in which it affects sound, through patterns of reflection and absorption. Just as a light source illuminates objects around it, sound sources allow us to hear the general shape and size of our physical environment. Because we are primarily oriented toward visual stimuli, it may take some dedicated and consistent effort to focus our awareness in the aural domain. As anyone who works in the field of audio engineering knows, the effort it takes to focus our aural awareness is well worth the satisfaction in acquiring critical listening skills. Although simple in concept, the practice of focusing attention on what is heard in a structured and organized way is challenging to accomplish in a consistent manner.

There are many situations outside of audio production in which listening skills can be developed. For instance, walking by a construction site, impulsive sounds such as hammering may be heard. Echoes—the result of those initial impulses reflecting from nearby building exteriors—can also be heard a short time later. The timing, location, and amplitude of echoes provide us with information about nearby buildings, including approximate distances to them.

Listening in a large concert hall, we notice that sound continues to linger on and slowly fade out after a source has stopped sounding. The gradual decaying of sound in a large acoustic space is referred to as *reverberation*. Sound in a concert hall can be enveloping because it seems to be coming from all directions, and sound produced on stage combines with reverberant sound arriving from all directions.

In a completely different location such as a carpeted living room, a musical instrument will sound noticeably different compared with the same instrument played in a concert hall. Physical characteristics such as dimensions and surface treatments of a living room determine that its acoustical characteristics will be markedly different than those of a concert hall; the reverberation time will be significantly shorter in a living room. The relatively close proximity of walls will reflect sound back toward a listener within milliseconds of the arrival of direct sound and at nearly the same amplitude. This small difference in time of arrival and near-equal amplitude of direct and reflected sound at the ears of a listener creates a change in the frequency content of the sound that is heard, due to a filtering of the sound known as *comb filtering*. Floor covering can also influence spectral balance: a carpeted floor will absorb some high frequencies, and a wood floor will reflect high frequencies.

When observing the surrounding sonic landscape, the listener may want to consider questions such as the following:

- What sounds are present at any given moment?
- Besides the more obvious sounds, are there any constant, steady-state, sustained sounds, such as air handling noise or lights humming, that are usually ignored?
- Where is each sound located? Are the locations clear and distinct, or diffuse and ambiguous?
- How far away are the sound sources?
- How loud are they?
- What is the character of the acoustic space? Are there any echoes? What is the reverberation decay time?

It can be informative to aurally analyze recorded music heard at any time whether it is in a store, club, restaurant, or elevator. It is useful to think about additional questions in such situations:

- How is the timbre of the sound affected by the system and environment through which it is presented?
- Are all of the elements of the sound clearly audible? If they are not, what elements are difficult to hear and which ones are most prominent?
- If the music is familiar, does the balance seem the same as what has been heard in other listening situations?

Active listening is critical in audio engineering, and we can take advantage of times when we are not specifically working on an audio project to heighten our awareness of the auditory landscape and practice our critical listening skills. Walking down the street, sitting in a café, and attending a live music concert all offer opportunities for us to hone our listening skills and thus improve our work with audio. For further study of some of these ideas, see Blesser and Salter's 2006 book *Spaces Speak, Are You Listening?*, where they expand upon listening to acoustic spaces in a detailed exploration of aural architecture.

Audio engineers are concerned with capturing, mixing, and shaping sound. Whether recording acoustic sound, such as from acoustic musical instruments playing in a live acoustic space, or creating electronic sounds in a digital medium, one of an engineer's goals is to shape sound so that it is most appropriate for reproduction over loudspeakers and headphones and best communicates the intentions of a musical artist. An important aspect of sound recording that an engineer seeks to control is the relative balance of instruments or sound sources, whether through manipulation of recorded audio signals or through microphone and ensemble placement. How sound sources are mixed and balanced in a recording can have a tremendous effect on the musical feel of a composition. Musical and spectral balance is critical to the overall impact of a recording.

Through the process of shaping sound, no matter what equipment is being used or what the end goal is, the main focus of the engineer is simply to listen. Engineers need to constantly analyze what they hear to assess a track or a mix and to help make decisions about further adjustments to balance and processing. Listening is an active process, challenging the engineer to remain continuously aware of any subtle and not so subtle perceived characteristics, changes, and defects in an audio signal.

From the producer to the third assistant engineer, active listening is a priority for everyone involved in any audio production process. No matter your role, practice thinking about and listening for the following items on each recording project:

- *Timbre.* Is a particular microphone in the right place for a given application? Does it need to be equalized? Is the overall timbre of a mix appropriate?
- *Dynamics.* Are sound levels varying too much, or not enough? Can each sound source be heard throughout piece? Are there any moments when a sound source

gets lost or covered by other sounds? Is there any sound source that is overpowering others?

- *Overall balance.* Does the balance of musical instruments and other sound sources make sense for the music? Or is there too much of one component and not enough of another?
- *Distortion/clipping.* Is any signal level too high, causing distortion?
- *Extraneous noise.* Is there a buzz or hum from a bad cable or connection or ground problem?
- *Space.* Is the reverb/delay/echo right?
- *Panning.* How is the left/right balance of the mix coming out of the loudspeakers?

1.1 What Is Technical Ear Training?

Just as musical ear training or *solfège* is an integral part of musical training, technical ear training is necessary for all who work in audio, whether in a recording studio, in live sound reinforcement, or in audio hardware/software development. Technical ear training is a type of *perceptual learning* focused on timbral, dynamic, and spatial attributes of sound as they relate to audio recording and production. In other words, heightened listening skills can be developed allowing an engineer to analyze and rely on auditory perceptions in a more concrete and consistent way. As Eleanor Gibson wrote, perceptual learning refers to "an increase in the ability to extract information from the environment, as a result of experience and practice with stimulation coming from it" (Gibson, 1969). This is not a new idea, and through years of working with audio, recording engineers generally develop strong critical listening skills. By increasing attention on specific types of sounds and comparing successively smaller differences between sounds, engineers can learn to differentiate among features of sounds. When two listeners, one expert and one novice, with identical hearing ability are presented with identical audio signals, an expert listener will likely be able to identify specific features of the audio that a novice listener will not. Through focused practice, a novice engineer can eventually learn to identify sounds and sound qualities that were originally indistinguishable.

A subset of technical ear training includes "timbral" ear training that focuses on the timbre of sound. One of the goals of pursuing this type of training is to become more adept at distinguishing and analyzing a variety of timbres. Timbre is typically defined as that characteristic of sound other than pitch or loudness, which allows a listener to distinguish two or more sounds. Timbre is a multidimensional attribute of sound and depends on a number of physical factors such as the following:

- *Spectral content.* All frequencies present in a sound.
- *Spectral balance.* The relative balance of individual frequencies or frequency ranges.
- *Amplitude envelope.* Primarily the attack (or onset) and decay time of the overall sound, but also that of individual overtones.

A person without specific training in audio or music can easily distinguish between the sound of a trumpet and a violin even if both are playing the same pitch at the same loudness—the two instruments *sound* different. In the world of recorded sound, engineers are often working with much more subtle differences in timbre that are not at all obvious to a casual listener. For instance, an engineer may be comparing the sound of two different microphone preamplifiers or two digital audio sampling rates. At this level of subtlety, a novice listener may hear no difference at all, but it is the experienced engineer's responsibility to be able to make decisions based on such subtle details.

Technical ear training focuses on the features, characteristics, and sonic artifacts that are produced by various types of signal processing commonly used in audio engineering, such as the following:

- Equalization and filtering
- Reverberation and delay
- Dynamics processing
- Characteristics of the stereo image

It also focuses on unwanted or unintended features, characteristics, and sonic artifacts that may be produced through faulty equipment, particular equipment connections, or parameter settings on equipment such as noise, hum or buzz, and unintentional nonlinear distortion.

Through concentrated and focused listening, an engineer should be able to identify sonic features that can positively or negatively impact a final audio mix and know how subjective impressions of timbre relate to physical control parameters. The ability to quickly focus on subtle details of sound and make decisions about them is the primary goal of an engineer.

The process of sound recording has had a profound effect on the development of music since the middle of the twentieth century. Music has been transformed from an art form that could be heard only through live performance into one where a recorded performance can be heard over and over again via a storage medium and playback system. Sound recordings can simply document a musical performance, or they may play a more active role in applying specific signal processing and timbral sculpting to recorded sounds. With a sound recording we are creating a virtual sound stage between our loudspeakers, in which instrumental and vocal sounds are located. Within this virtual stage recording engineers can place each instrument and sound.

With technical ear training, we are focusing not only on hearing specific features of sound but also on identifying specific sonic characteristics and types of processing that cause a characteristic to be audible. It is one thing to be able to know that a difference exists between an equalized and nonequalized recording, but it is quite another to be able to name the specific alteration in terms of center frequency, Q, and gain. Just as experts in visual art and graphic design can identify subtle shades and hues of color by name, audio professionals should be able to do the same in the auditory domain.

Sound engineers, hardware and software designers, and developers of the latest perceptual encoders all rely on critical listening skills to help make decisions about a variety of characteristics of sound and sound processing. Many characteristics can be measured in objective ways with test equipment and test signals such as pink noise and sine tones. Unfortunately, these objective measures do not always give a complete picture of how equipment will sound to human ears using musical signals. Some researchers such as Geddes and Lee (2003) have pointed out that high levels of measured nonlinear distortion in a device can be less perceptible to listeners than low levels of measured distortion, depending on the nature of the distortion and the testing methods employed. The opposite can also be true, in that low levels of measured distortion can be perceived strongly by listeners.

This type of situation can be true for other audio specifications such as frequency response. Listeners may prefer a loudspeaker that does not have a flat frequency response over one that does because frequency response is only one objective measurement of the total sound produced by a loudspeaker. In other areas of audio product design, the final tuning of software algorithms and hardware designs is often done by ear by expert listeners. Thus, physical measurements cannot be solely relied upon, and often it is auditory perceptions that determine the final verdict on sound quality.

Professionals who work with recorded sound on a daily basis understand the need to hear subtle changes in sound. It is important to know not only how these changes came about but also ways in which to use the tools available to remedy any problematic characteristics.

1.1.1 Isomorphic Mapping

Professionals who work with recorded sound on a daily basis understand the need to hear subtle changes in sound. It is important to know not only how these changes came about but also ways in which to use the tools available to remedy any problematic characteristics. One of the primary goals of this book is to facilitate isomorphic mapping of technical and engineering parameters to perceptual attributes; to assist in the linking of auditory perceptions with control of physical properties of audio signals.

With audio recording technology, engineers have control over technical parameters that correspond to physical attributes of an audio signal, but often it is not clear to the novice how to map a perceived sensation to the control of objective parameters of the sound. A parametric equalizer, for instance, usually allows us to control frequency, gain, and Q. These physical attributes as they are labeled on a device have no natural or obvious correlation to perceptual attributes of an audio signal, and yet engineers engage them to affect a listener's perception of a signal. How does an engineer know what a 6-dB boost at 315 Hz with a Q of 2 sounds like? Without extensive experience with equalizers, these numbers will have little meaning in terms of how they affect the perceived timbre of a sound.

There exists an *isomorphism* between audio equipment that are typically used to make a recording and the type of sound an engineer hears and wishes to obtain. An engineer can form mental links between particular features of sound quality and specific types of signal processing or equipment. For example, a novice audio engineer may understand what the term *compression ratio* means in theory, but the engineer may not know how to adjust that parameter on a compressor to effectively alter the sound or may not fully understand how sound is changed when that parameter is adjusted. One important component of teaching audio engineering is to illustrate the mapping between engineering concepts and their respective effect on the sound being heard. To teach these concepts requires the use of audio examples and also specific training for each type of processing. Ear training is equally important as knowing the functionality of equipment on hand. Letowski, in his article "Development of Technical Listening Skills: Timbre Solfeggio" (1985), originally coined the term *timbre solfeggio* to designate training that has similarities to musical aural training but is focused on spectral balance or timbre.

If an engineer uses words such as *bright* or *muddy* to describe the quality of a sound, it is not clear exactly what physical characteristics are responsible for a particular subjective quality; it could be specific frequencies, resonances, dynamics processing, artificial reverberation, or some combination of all of these and more. There is no label on an equalizer indicating how to affect these subjective parameters. Likewise, subjective descriptions by their nature are not always consistent from person to person or across situations. A "bright"-sounding snare drum may mean excessive energy around 4 to 8 kHz in one situation or a deficiency around 125 Hz in another. It is difficult to be precise with subjective descriptions of sound, but ambiguity can be reduced if everyone agrees on the exact meaning of the adjectives being used.

Continuing with the example, an equalizer requires that a specific frequency be chosen to boost or cut, but a verbal adjective chosen to describe a sound may only give an imprecise indication that the actual frequency is in the low-, mid-, or high-frequency range. It is critical to develop an internal map of specific frequencies to perceptual attributes of a signal, and what a boost or cut at specific frequencies sounds like. With practice it is possible to learn to estimate the frequency of a deficiency or surplus of energy in the power spectrum of an audio signal and then fine-tune it by ear.

Through years of practice, professional audio engineers develop methods to translate between their perceived auditory sensations and the technical parameters that they can control with the equipment available to them. They also develop a highly tuned awareness of subtle details present in sound recordings. Although there may not be a common language among recording engineers to describe specific auditory stimuli, those engineers working at a very high level have devised their own personal translation between the sound they hear and imagine, and the signal processing tools available. Comparing audiological exams between professional and novice engineers would likely not demonstrate superior hearing abilities in the professionals from a clinical, objective standpoint. Something else is going on: professionals are more advanced in their ability to focus on sound.

A recording engineer should ideally have as much command of a recording studio and its associated signal processing capability as a professional musician has command of her instrument. A professional violinist knows precisely when and where to place her fingers on the strings and precisely what effect each bow movement will have on the sound produced. There are an intimate knowledge and anticipation of a sound even before it is produced. An audio engineer should have this same level of knowledge and sensitivity of sound processing and shaping before reaching for an effects processor parameter, fader position, or microphone model. It is important to know what a 3-dB boost at 4 kHz or an increase in compression ratio is going to sound like even before it is applied to an audio signal. There will always be times when a unique combination of signal processing and equipment choices will not be immediately apparent, but it is highly inefficient for an engineer to be continuously guessing what the standard types of studio signal processing will sound like. By knowing ahead of time what a particular parameter change will have on the sound quality of a recorded signal, an engineer can work more efficiently and effectively. Working at such a high level, an engineer is able to respond to sound quality very quickly, similar to the speed in which musicians respond to each other in an ensemble.

A recording studio can be considered as a musical instrument that is "played" by a recording engineer and producer. An engineer has direct input and influence on the artistic outcome of any music recording in which she is involved. By adjusting balances and shaping spectra, an engineer focuses the sonic scene for listeners, guiding them aurally to a musically satisfying experience that expresses the intentions of the musical artist.

1.1.2 Increasing Awareness

The second goal of technical ear training is to increase our awareness of subtle details of sound and develop our ability to discern and identify by ear minute changes in physical parameters. An experienced recording engineer or producer can focus her attention on details of sound that may not be apparent to an untrained listener. Often, the process of making a recording from start to finish is built on hundreds, if not thousands, of decisions about technical aspects of sound quality and timbre. Each decision contributes to a finished project and influences other choices. These decisions encompass a wide range of options and level of subtlety but typically include:

- Microphone model, location, and orientation for each instrument being recorded.
- Preamplifier model and gain settings for each microphone.
- Recording level—which must be set high enough to reduce noise and quantization error, and low enough to avoid overloading a gain stage.

- Equalizer model and specific equalization parameter settings for each microphone signal.
- Noise—which can take many forms but in general is any sound that is not intended to be part of a recording. Examples include clicks/pops produced by analog or digital electronics, tape hiss, quantization error, air handling noise (which can be in the form of a low rumble and therefore not immediately apparent), external and environmental sounds such as traffic and subways, 50- or 60-Hz buzz or hum.
- Timbral quality—primarily frequency content and spectral balance. Every analog component from the microphone to the input of the recording device, as well as every stage of analog to digital conversion and re-quantization will have some effect on the timbral quality of audio.
- Dynamic range and dynamics processing—sound, musical or otherwise, will have a certain range from loud (fortissimo) to soft (pianissimo), and this range can be altered through dynamics processing, such as compressors and expanders.
- Balancing or mixing levels of recorded microphone signals.
- Spatial characteristics—includes reverberation, echo, reflections, delays, as well as panning and positioning of sound sources within the stereo or surround image.

An engineer makes decisions concerning these and other technical parameters that affect the perceived audio quality and timbre of an audio signal.

It may be tempting to consider these subtle changes as insignificant, but because they are added together to form a coherent whole, the cumulative effect makes each stage critical to a finished project. Whether it is the quality of each component of a sound system or each decision made at every stage of a recording project, the additive effect is noteworthy and substantial. Choices made early in a project which degrade sound quality cannot be reversed later in a project. Audio problems cannot be fixed in the mix and, as such, engineers must be listening intently to each and every decision about signal path and processing that is made. When listening at such a focused level, an engineer can respond to sound quality and timbre quickly and in the moment, hearing potential problems which may come back to haunt a project at a later stage. To use an analogy, painters use specific paint colors and brush strokes in subtle ways that combine to produce powerful finished images. In a related way, recording engineers must be able to hear and focus on specific sonic characteristics that, when taken as a whole, combine, blend, and support one another to create more powerful, meaningful final mixtures of sounds.

1.1.3 Increasing Speed of Detection

Finally, the third goal is to increase the speed with which we can identify and decide on appropriate engineering parameters to change. A recording and mixing session can occupy large amounts of time, within which hundreds of subtle and not-so-subtle adjustments can be made. The faster an engineer can home in on any sonic characteristics that may need to be changed, the more effective a given period of time will be. The ability to make quick judgments about sound quality is paramount during recording and mixing sessions. For example, during a recording session, valuable time can be consumed while comparing and changing microphones.

It is anticipated that increased sensitivity in one area of critical listening (such as equalization) will facilitate increased awareness and sensitivity in other areas (such as compression and reverberation) as a result of overall improved listening skills. Because a significant portion of audio engineering—recording, mixing, mastering—is an art in which there are no correct answers, this book does not provide advice on the "best" equalization, compression, or reverberation settings for different situations. What may be the perfect equalization for an instrument in one situation may not be suitable for another. What this book attempts to do, however, is guide the reader in the development of listening skills that then assist in identifying problematic areas in sound quality. A novice engineer may not realize when there is a problem with sound quality or may have some idea that there is a problem but may not be able to identify it specifically or know how to solve it. Highly developed critical listening skills help an engineer identify characteristics of timbre and sound quality quickly and efficiently.

The standard signal processing types include equalization (parametric, graphic, and filters), compression/limiting, expansion/gating, reverberation, delay, chorus, flanging, and gain changes. Within each of these categories of signal processing, numerous makes and models are available at various price ranges and levels of quality. If we consider compressors for a moment, we know that various compressor makes/models perform the same basic function—they make loud sounds quieter. Most compressor models have common functionalities that give them similar general sonic characteristics, but the exact way in which they perform gain reduction varies from model to model. Differences in the analog electronics or digital signal processing algorithms among compressors create a variety of sonic results, and each make and model will have a unique sound. Through the experience of listening, engineers learn that there are variations in sound quality between different makes and models, and they will choose a certain model because of its specific sound quality.

It is common to find software plug-in versions of many analog signal processing devices. Often the screen image of a plug-in modeling an analog device will be nearly identical to the faceplate of the device. Sometimes, because the two devices look identical, it may be tempting to think that they also sound identical. Unfortunately, they do not always sound alike but it is possible to be fooled into thinking the sound is replicated as perfectly as well as the visual representation of the device. Usually the best option is to listen and determine by ear if the two sound as similar as they look. There is not always a direct translation between analog electronics and the computer code that performs the equivalent digital signal processing, and there are various ways to create models of analog circuits; thus we have differences in sound quality.

Although each signal processing model has a unique sound, it is possible to transfer knowledge of one model to another and be able to use an unknown model effectively after a short period of listening. Just as pianists must adjust to each new piano that they encounter, engineers must adjust to the subtle and not-so-subtle differences between pieces of equipment that perform a given function.

1.2 Shaping Sounds

Not only can music recordings be recognized by their musical melodies, harmonies, and structure, they can also be recognized by the timbres of the instruments created in the recording process. Sometimes timbre is the most identifying feature of a recording. In recorded music, an engineer and producer shape sounds that are captured to best suit a musical composition. The molding of timbre has become incredibly important in recorded music, and in his book *The Producer as Composer: Shaping the Sounds of Popular Music* (2005), Moorefield outlines how recording and sound processing equipment contribute to the compositional process. Timbre has become such an important factor in recorded music that it can be used to identify a song before musical tonality or melody can have time to develop sufficiently. In their article titled "Name That Tune: Identifying Popular Recordings from Brief Excerpts," Schellenberg et al. (1999) found that listeners could correctly identify pieces of music when presented with excerpts of only a tenth of a second in length. Popular music radio stations are known to challenge listeners by playing a short excerpt (typically less than a second) from a well-known recording and inviting listeners to call in and identify the song title and artist. Such excerpts are too short to indicate the harmonic or melodic progression of the music. Listeners rely on the timbre or "mix" of sonic features to make a correct identification. Levitin, in *This Is Your Brain on Music* (2006), also illustrates the importance of timbre in recorded sound and reports that "Paul Simon thinks in terms of timbre; it is the first thing he listens for in his music and the music of others" (page 152).

One effect that the recording studio has had on music is that it has helped musicians and composers create sonic landscapes that are impossible to realize acoustically. Sounds and sound images that could not have been produced acoustically are most evident in electroacoustic and electronic music in which sounds originate from purely electronic or digital sources rather than through a conventional musical instrument's vibrating string, membrane, or airflow. Nonetheless, recordings of purely acoustic musical instruments can be significantly altered with standard recording studio processing equipment and plug-ins. Electronic processing of the spectral, spatial, and dynamic properties of recorded sound all alter a sound source's original properties, creating new sounds that might not exist as purely acoustic events.

In the process of recording and mixing, an engineer can manipulate any number of parameters, depending on the complexity of a mix. Many of the parameters that are adjusted during a mix are interrelated, such that by altering one track the perception of other tracks is also influenced. The level of each instrument can affect the entire feel or focus of a mix, and an engineer and producer may spend countless hours adjusting levels—down to increments of a quarter of a decibel—to create the right balance. As an example, a slight increase in the level of an electric bass may have a significant impact on the sound and musical feel of a kick drum or even an entire mix as a whole. Each parameter change applied to an audio track, whether it is level (gain), compression, reverberation, or equalization, can have an effect on the perception of other individual instruments and the music as a whole. Because of this interrelation between components of a mix, an engineer may wish to make small, incremental changes and adjustments, gradually building and sculpting a mix.

At this point, it is still not possible to measure all perceived audio qualities with the physical measurement tools currently available. For example, the development of perceptual coding schemes such as MPEG-1 Layer 3, more commonly known as MP3, has required the use of expert listening panels to identify sonic artifacts and deficiencies produced by data reduction processes. Because perceptual coding relies on psychoacoustic models to remove components of a sound recording that are deemed inaudible, the only reliable test for this type of processing is the human ear. Small panels of trained listeners are more effective than large samples of the general population because they can provide consistent judgments about sound and they can focus on the subtlest aspects of a sound recording.

Studies, such as those by Quesnel (2001) and Olive (1994, 2001), provide strong evidence that training people to hear specific attributes of reproduced sound makes a significant difference in their ability to consistently and reliably recognize features of sound, and it also increases the speed with which they can correctly identify these features. Listeners who have completed systematic timbral ear training are able to work with audio more productively and effectively.

1.3 Sound Reproduction System Configurations

Before examining critical listening techniques and philosophies more closely, it is important to outline what some of the most common sound reproduction systems look like. Recording engineers are primarily concerned with sound reproduced over loudspeakers, but there is also benefit to analyzing acoustic sound sources, as we will discuss in Chapter 7.

1.3.1 Monaural: Single-Channel Sound Reproduction

A single channel of audio reproduced over a loudspeaker is typically called monaural or mono (Fig. 1.1). Even if there is more than one loudspeaker, it is still considered monaural if all loudspeakers are producing exactly the same audio signal. The earliest sound recording, reproduction, and broadcast

Figure 1.1 Monaural or single-channel listening.

systems used only one channel of audio, and although this method is not as common as it once was, we still encounter situations where it is used. Mono sound reproduction creates some restrictions for a recording engineer, but it is often this type of system that loudspeaker manufacturers use for subjective evaluation and testing of their products.

1.3.2 Stereo: Two-Channel Sound Reproduction

Evolving from monaural systems two-channel reproduction systems, or stereo, allow sound engineers greater freedom in terms of sound source location, panning, width, and spaciousness. Stereo is the primary configuration for sound reproduction whether using speakers or headphones. Figure 1.2 shows the ideal listener and loudspeaker locations for two-channel stereo.

1.3.3 Headphones

Headphone listening with two-channel audio has advantages and disadvantages with respect to loudspeakers. With modestly priced headphones (relative to the price of equivalent quality loudspeakers), it is possible to achieve high-quality sound reproduction. Good-quality headphones can offer

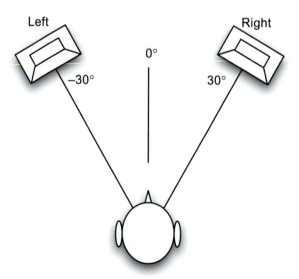

Figure 1.2 Ideal two-channel stereo listening placement.

more clarity and detail than loudspeakers, partly because they are not subject to the acoustical effects of listening rooms such as early reflections and room modes. Headphones are also portable and can be easily taken to other locations where loudspeaker characteristics and room acoustics may be unfamiliar to an engineer.

The main disadvantage of headphones is that they create in-head localization for mono sound sources. That is, center-panned, mono sounds are perceived to be originating somewhere between the ears because the sound is being transmitted directly into the ears without first bending around or reflecting off the head, torso, and outer ear. To avoid in-head localization, audio signals would need to be filtered with what are known as head-related transfer functions (HRTF). Simply put, HRTFs specify filtering because of the presence of outer ears (pinnae), head, and shoulders, as well as interaural time differences and interaural amplitude differences for a given sound source location. Each location in space (elevation and azimuth) has a unique HRTF, and usually many locations in space are sampled when measuring HRTFs. It is also worth noting that each person has a unique HRTF based on the unique shape of the outer ear, head, and upper torso. HRTF processing has a number of drawbacks such as a negative effect on the sound quality and spectral balance and the fact that there is no universal HRTF that works perfectly for everyone.

1.3.4 Headphone Recommendations

As of this writing, there are a number of fine headphones on the market that are perfectly suitable for technical ear training. Before purchasing headphones, the reader is encouraged to listen to as many different models as possible. By comparing the sound of different headphones using music recordings that are familiar it is possible to get a better sense of the strengths and weaknesses of each model. There is no perfect headphone, and each model will have a slightly different sound. Because not all readers are near retail stores that stock high-quality headphones, some suggestions are made here at varying price points:

- *Audio-Technica ATH-M50.* This model is a closed design, meaning that it blocks out a substantial amount of external or background sound.

- *Beyerdynamic DT770 Pro.* This model is also a closed back design with a comfortable circumaural fit.
- *Grado.* There are a number of models in the Grado headphone line and all are supra-aural designs, meaning that they rest right on the ear, as opposed to being circumaural, which surround the ear. Furthermore, they are all open headphones meaning that they do not block outside sound and thus might not be appropriate for listening in environments where there is significant background noise. Grado headphones are an excellent value for the money, especially for the lower-end models, despite the fact that they are not the most comfortable headphones available.
- *Sennheiser HD 600 and HD 650.* Both of these models are open design and on the higher end of the price range for headphones. They are also circumaural in design, making them comfortable to wear.
- *Sony MDR 7506 and 7509.* These models from Sony have become somewhat of an industry standard for studio monitoring.

1.3.5 Surround: Multichannel Sound Reproduction

Sound reproduced over more than two loudspeakers is known as *multichannel, surround, ambisonic,* or more specific notations indicating numbers of channels, such as *5.1, 7.1, 3/2 channel,* and *quadraphonic.* Surround audio for music-only applications has had limited popularity and is still not as popular as stereo reproduction. On the other hand, surround soundtracks for film and television are common in cinemas and are becoming more common in home systems.

There are many suggestions and philosophies on the exact number and layout of loudspeakers for surround reproduction systems, but the most widely accepted configuration among audio researchers is from the International Telecommunications Union (ITU), which recommends a five-channel loudspeaker layout as shown in Figure 1.3. Users of the ITU-recommended configuration generally also make use of an optional subwoofer or low-frequency effects (LFE) channel known as the *.1* channel, which reproduces only low frequencies, typically below 120 Hz.

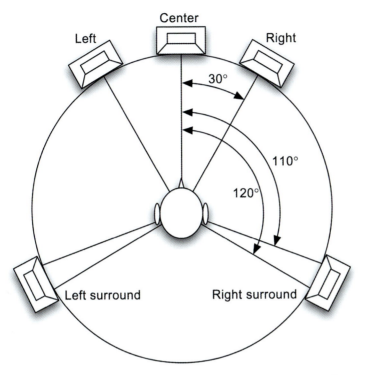

Figure 1.3 Ideal five-channel surround listening placement according to the ITU-R BS.775-1 recommendations (ITU-R, 1994), with the listener equidistant from all five loudspeakers.

With multichannel sound systems, there is much more freedom for sound source placement within the 360° horizontal plane than is possible with stereo. There are also more possibilities for convincing simulation of immersion within a virtual acoustic space. Feeding the appropriate signals to the appropriate channels can create a realistic sense of spaciousness and envelopment. As Bradley and Soulodre (1995) have demonstrated, listener envelopment (LEV) in a concert hall, a component of spatial impression, is primarily dependent on having strong lateral reflections arriving at the listener 80 ms or more after the direct sound.

There are also some challenges with respect to sound localization for certain areas within a multichannel listening area. Panning sources to either side (between 30° and 110°) produces sound images that are unstable and difficult to accurately localize. On the other hand, the presence of a center channel allows sounds to be locked into the center of the front sound image, no matter where a listener is located. When sources are panned to the center with only two loudspeakers in front (left and right), the perceived location of the image depends on a listener's location.

Summary

In this chapter we have explored active listening and its importance in recording projects as well as everyday life. By defining technical ear training, we also identified some goals toward which we are working through the book and software practice modules. We finished by giving a rough overview of the main sound reproduction systems. Next we will move on to more specific ideas and exercises focused on equalization.

2

SPECTRAL BALANCE AND EQUALIZATION

CHAPTER OUTLINE

Audio Production and Critical Listening. DOI: 10.1016/B978-0-240-81295-3.00002-2
Copyright © 2010 by Focal Press. Inc. All rights of reproduction in any form reserved.

Spectral balance refers to an audio signal's frequency content and the relative power of each frequency or frequency band across the audible range of frequencies, from 20 to 20,000 Hz. An audio signal with a flat spectral balance would represent all frequencies at the same relative amplitude. Often audio engineers describe the spectral balance of sound through equalization parameters, as the equalizer is the primary tool for altering the spectral balance of sound. An engineer can boost or cut specific frequencies or ranges of frequencies with an equalizer to bring out low-level details or to compensate for unwanted resonances.

In the context of sound recording and production, a flat spectral balance is more likely to mean that the entire range of frequencies in a recording of a sound source is represented appropriately for a given recording project. Yet it is not always clear what we mean by representing all frequencies "appropriately." Does it mean that we want recordings of musical instruments to sound identical to how they sound acoustically? Is that possible or even desirable? In classical music recording, engineers usually strive for some similarity to live performances, but in most other genres of music, engineers are creating sound images that do not exist in a live performance situation. Sounds and timbres are created and shaped in the recording studio and digital audio workstation, making it possible to take recorded sound in many possible artistic directions.

Although the equalizer is the main tool for directly altering spectral balance, almost every electronic device through which audio passes alters the spectral balance of an audio signal to a greater or lesser extent. Sometimes this alteration of frequency content is necessary and completely intentional, such as with the use of equalizers and filters. Other times a change in the spectral balance is much more subtle or nearly imperceptible, as in that caused by different types of microphone preamplifiers. Vintage audio equipment is often sought after because of unique and pleasing alterations to the spectral balance of an audio signal. Changes in spectral balance are sometimes caused by distortion, which results in harmonics being added to an audio signal. Audio engineers should be able to hear how each piece of audio equipment is altering the spectral content of their audio signals to shape the timbre of each sound so that it is most appropriate for a given situation. The ability to distinguish subtle yet critical aspects of sound quality comes through the experience of listening to various types of audio processing and forming mental links between what one hears and what parameters can be controlled in an audio signal. In essence, experienced audio professionals are like human spectral analyzers because of their ability to identify and characterize the frequency balance of reproduced sound.

Aside from the use of equalizers, spectral balance can also be altered to a certain extent through dynamics processing, which changes the amplitude envelope of a signal and, by consequence, its frequency content, and through mixing a signal with a delayed version of itself, which can produce comb filtering. Although both of these methods influence spectral balance, we are going to focus on signal processing devices whose primary function is to alter the frequency content of a signal.

An engineer seeks the equalization and spectral balance that is best suited to whatever music is being recorded. For instance, the spectral balance appropriate for a jazz drum kit recording will likely be different from that for a rock drum recording, and an experienced recording engineer, upon listening to two such audio samples, understands and can identify specific timbral differences between them.

To determine the equalization or spectral balance that best suits a given recording situation, an engineer must have well-developed listening skills with regard to frequency content and its relationship to physical parameters of equalization: frequency, gain, and Q. Each recording situation calls for specific engineering choices, and there are rarely any general recommendations for equalization that are applicable across multiple situations. When approaching a recording project, an engineer should be familiar with existing recordings of a similar musical genre or have some idea of the timbral goals for a project to inform the decision process during production.

An engineer monitors the spectral balance of individual microphone signals as well as the overall spectral balance of multiple, combined microphone signals at each stage in a recording project. It is possible to use a real-time spectral analyzer to get some idea of the frequency content and balance of an audio signal. A novice engineer may wish to employ a real-time spectral analyzer to visualize the frequency content of an audio signal and apply equalization based on what he sees. Professional recording and mixing engineers do not usually measure the power spectrum of a music signal but instead rely on their auditory perception of the spectral balance over the course of a piece of music.[1] Unfortunately, real-time analyzers do not give a clear enough picture of the frequency content of a music recording to rely on it for decisions about how to apply equalization to a music signal. Furthermore, there is no clear indication of what the spectral plot "should" look like because there is no objective reference.

Music signals generally exhibit constant fluctuations, however large or small, in frequency and amplitude of each harmonic and overtone present. Because of the constantly changing nature of a typical music signal, it becomes difficult to get a clear reading of the amplitude of harmonics. Taking a snapshot of a spectral plot from a specific moment in time would be clearer visually, but it does not give a broad enough view of an audio signal's general spectral shape over time. The situation is complicated a bit more because with any objective spectral analysis there is a trade-off between time resolution and frequency resolution. With increases in time resolution, the frequency resolution decreases while the display of frequency response updates at such a fast rate that it is difficult to see details accurately while an audio signal is being played back. Thus, physical measures currently available are not appropriate for determining what equalization to apply to a music signal, and the auditory system must be relied upon for decisions about equalization.

---

[1]Live sound engineers, on the other hand, who are tuning a sound system for a live music performance will often use real-time spectral analyzers. The difference is that they have a reference, which is often pink noise or a recording, and the analyzer compares the spectrum of the original audio signal (a known, objective reference) to the output of the loudspeakers. The goal in this situation is a bit different from what it is for recording and mixing because a live sound engineer is adjusting the frequency response of a sound system so that the input reference and the system output spectral balances are as similar as possible.

2.1 Shaping Spectral Balance

2.1.1 Equalization

In its most basic characterization, spectral balance can refer to the relative balance of bass and treble, what can be controlled with basic tone controls on a consumer sound system. Typically during the process of recording an acoustic musical instrument, an engineer can have direct control over the spectral balance of recorded sound, whether a single audio track or a mix of tracks, through a number of different methods. Aside from an equalizer, the most direct tool for altering frequency balance, there are other methods available to control the spectral balance of a recorded audio track, as well as indirect factors that influence perceived spectral balance. In this section we discuss how engineers can directly alter the spectral balance of recorded sound, as well as ways in which spectral balance can be indirectly altered during sound reproduction.

The most obviously deliberate method of shaping the spectral balance of an audio signal is accomplished with an equalizer or filter, a device specifically designed to change the amplitude of selected frequencies. Equalizers can be used to reduce particular frequency resonances in a sound recording, since they can mask other frequency components of a recorded sound and prevent the listener from hearing the truest sound of an instrument. Besides helping to remove problematic frequency regions, equalizers can also be used to accentuate or boost certain frequency bands to highlight characteristics of an instrument or mix. There is a significant amount of art in the use of equalization, whether for a loudspeaker system or a recording, and an engineer must rely on what is being heard to make decisions about its application. The precise choice of frequency, gain, and Q is critical to the successful use of equalization, and the ear is the final judge of the appropriateness of an equalizer setting.

2.1.2 Microphone Choice and Placement

Another method of altering the spectral balance of an audio signal is through a microphone. The choice of microphone type and model has a significant effect on the spectral balance of any sound being recorded, as each make and model of microphone has a unique frequency response because of internal electronics and physical construction. Microphones are analogous to filters or lenses on a camera; microphones affect not only the overall frequency content but also the perspective and clarity of the sound being "picked up." Some microphone models offer a frequency response that is very close to flat, whereas others are chosen because they are decidedly not flat in their frequency response. Engineers often choose microphones because of their unique frequency responses and how the frequency response relates to the sound source being recorded.

During a beginning of a recording session, a recording engineer and producer compare the sounds of microphones to decide which ones to use for a recording. By listening to different microphones while musicians are performing, they can decide which microphones have the sonic characteristics that are most appropriate for a given situation. The choice would take into account the characteristics of a musician's instrument or voice, the space in which they are recording, and any blending that may need to occur with other instruments/ voices that are also being picked up by the microphone.

Besides a microphone's frequency response, its physical orientation and location relative to a sound source also directly affect the spectral balance of the audio signal as other factors come into play, such as the polar response of the microphone, the radiation patterns of a sound source, and the ratio of direct sound to reverberant sound at a given location within an acoustic space. The location of a microphone in relation to a musical instrument can have a direct and clear effect on the spectral balance of the sound picked up. Sound radiated from a musical instrument does not have the same spectral balance in all directions. As an example, sound emanating directly in front of a trumpet bell will contain a much higher level of high-frequency harmonics than sound to the side of the trumpet. An engineer can affect the frequency response of a recorded trumpet sound by simply changing the location of a microphone relative to the instrument. In this example, having the player aim the trumpet bell slightly above or below a microphone will result in a slightly darker sound than when the trumpet is aimed directly at a microphone.

Beyond the complex sound radiation characteristics of musical instruments, microphones themselves do not generally have the same frequency response for all angles of sound incidence. Even omnidirectional microphones, which are generally considered to have the best off-axis response, have some variation in their frequency response across various angles of sound incidence. Simply changing the angle of orientation of a microphone can alter the spectral balance of a sound source being recorded.

Directional microphones—such as cardioid and bidirectional polar patterns—produce an increased level of low frequencies when placed close to a sound source, in a phenomenon known as *proximity effect* or *bass tip-up*. A microphone's response varies in the low-frequency range according to its distance to a sound source, within a range of about 1 m. It is important to be aware of changes in low-frequency response as a result of changes in a musician's distance from a microphone. This effect can be used to advantage to achieve prominent low frequencies when close miking a bass drum, for instance.

2.1.3 Indirect Factors Affecting Spectral Balance

When working on shaping the spectral balance of a track or mix, there are a few factors that will have an indirect influence on this process. Because there is no direct connection between the auditory processing center of the brain and digital audio data or analog magnetic tape, engineers need to keep in mind that audio signals are altered in the transmission path between a recorder and the brain. Three main factors influence our perception of the spectral balance of an audio signal in our studio control room:

- Monitors/loudspeakers
- Room acoustics
- Sound levels

Figure 2.1 illustrates the path of an audio signal from electrical to acoustical energy, highlighting three of the main modifiers of spectral balance.

2.1.3.1 Monitors and Loudspeakers

Monitors and loudspeakers are like windows through which engineers perceive and therefore make decisions upon recorded audio signals. Although monitors do not have a direct effect on the spectral balance of signals sent to a

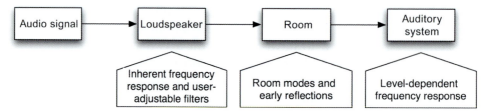

Figure 2.1 The signal path showing the transmission of an audio signal as an electrical signal to a loudspeaker where it is converted to an acoustic signal, modified by a listening room, and finally received by the ear and processed by the auditory system. Each stage highlights factors that influence a signal's spectral balance—both physical and perceptual—through the path.

recorder, each type and model of monitor and loudspeaker offers a unique frequency response. Because engineers rely on monitors to judge the spectral balance of audio signals, the frequency and power response of monitors can indirectly alter the spectral balance of the audio signals. Listening to a recording through monitors that have a weak low-frequency response, an engineer may have a tendency to boost the low frequencies in the recorded audio signal. It is common for engineers to check a mix on three or more different sets of monitors and headphones to form a more accurate conception of what the true spectral balance of the audio signal is. Each model of loudspeaker is going to give a slightly different impression, and by listening to a variety of monitors, engineers can find the best compromise. Beyond the inherent frequency response of a loudspeaker, almost all active loudspeakers include built-in user-adjustable filters—such as high- and low-frequency shelving filters—that can compensate for such things as low-frequency buildup when monitors are placed close to a wall. So any decisions made about spectral balance will be influenced by the cumulative effect of a speaker's inherent frequency response added to any filtering applied by the user.

Real-time analyzers can provide some indication of the frequency response of a loudspeaker within a room, and equalizers can be used to adjust a response until it is nearly flat. One important point to keep in mind that unless frequency response is being measured in an anechoic chamber, the response that is presented is not purely that of the loudspeaker but will also include room resonances and reflections. Any type of objective frequency response measurements performed in a listening room or studio should be averaged over different locations in the listening area. As we will discuss in the next section, frequency resonances in a room are prominent in some locations and less so in others. By measuring frequency response from different locations, we average the effect of location-dependent resonances.

2.1.3.2 Control Room and Listening Room Acoustics

The dimensions, volume, and surface treatments of the room in which an engineer monitors audio signals also have a direct effect on the audio being heard. Groups such as the International Telecommunications Union (ITU) have published recommendations on listening room acoustics and characteristics. Recommendation ITU-R BS.1116 (ITU-R, 1997) defines a number of physical and acoustical parameters that can be applied to a listening room to create an acoustically neutral room. It may seem upon initial thought that an anechoic room free of room modes and reflections would be ideal for listening because the room will essentially be "invisible" acoustically, but a reflection-free room does not give us a realistic environment that mirrors the type of room in which we typically hear music. Sound originating from loudspeakers propagates into a room, reflects off objects and walls, and combines with the sound propagating directly to the listener. Sound radiates mainly from the front of a loudspeaker especially for high frequencies, but most loudspeakers become more omnidirectional as frequency decreases. The primarily low-frequency sound that is radiated from the back and sides of a loudspeaker will be reflected back into the listening position by any wall that may be behind the loudspeaker. Regardless of the environment in which we are listening to reproduced sound, we hear not only the loudspeakers but also the room. In essence the loudspeakers and listening environment act as a filter, altering the sound we hear.

Room modes depend on a room's dimensions and influence the spectral balance of what is heard from loudspeakers in a room. Room modes are mostly problematic in the low-frequency range, typically below 300 Hz. Fundamental resonant frequencies that occur in one dimension (axial modes) have wavelengths that are two times the distance between parallel walls. Splaying or angling walls does not reduce room modes; instead the resonant frequencies are based on the average distance between opposing walls.

Because the amplitudes of room resonances vary according to location, it is important for an engineer to walk around and listen at different locations within a room. The listening position of a room may have a standing wave node at a particular frequency. Without realizing this low-frequency acoustical effect, a mixing engineer may boost the missing frequency with an equalizer, only to realize when listening at a different location in the room that the frequency boost is too great.

If a mixing studio is attached to an adjacent room that is available, engineers like to take a walk into the second room, leaving the adjoining door open, and audition a mix, now essentially filtered through two rooms. Listening to the balance of a mix from this new location, an engineer can learn which components of the balance change from this new perspective, which sounds remain prominent, and which ones get lost. It can be useful to focus on how well the vocals or lead instrument can be heard from a distant listening location.

Another common and useful way of working is to audition a mix on a second and possibly third set of speakers and headphones, because each set of speakers will tell us something different about the sound quality and mix balance. One loudspeaker set may be giving the impression that the reverberation is too loud, whereas another may sound like there is not enough bass. Among the available monitoring systems, a compromise can be found that one hopes will allow the final mix to sound relatively optimal on many other systems as well. Engineers often say that a mix "translates" well to describe how consistent a mix remains when auditioned on various types and sizes of loudspeakers. There can be enormous differences highlighted in a mix auditioned on different systems, depending on how the mix was made. One mark of a well-made recording is that it will translate well on a wide range of sound reproduction systems, from minisystems to large-scale loudspeaker systems.

2.1.3.3 Sound Levels and Spectral Balance

The sound level of a sound reproduction system plays a significant role in the perception of spectral balance. The well-known equal-loudness contours by Fletcher and Munson (1933) illustrate that not only does the human auditory system have a wide variation in its frequency response, but also that this response changes according to sound reproduction level. In general, the ear is less sensitive to low and high frequencies, but as sound level is increased, the ear becomes more sensitive to these same frequencies, relative to midfrequencies. If mixing at a high sound level—such as 100 dB average sound pressure level—and then suddenly the level is turned down much lower—to 55 dB SPL, for example—the perceived spectral balance will change. There will be a tendency to think that there are not enough low frequencies in the mix. It is useful to listen to a mix at various reproduction levels and find the best compromise in overall spectral balance, taking into account the frequency response differences of the human auditory system at different reproduction levels.

2.2 Types of Filters and Equalizers

Now that we have discussed ways to change spectral balance directly as well as factors that are responsible for altering our perception of reproduced sound, it is time to focus more specifically on equalizers. There are different types of equalizers and filters such as high-pass filters, low-pass filters, band-pass filters, graphic equalizers, and parametric equalizers, allowing various levels of control over spectral balance. Filters are those devices that remove a range or band of frequencies, above or below a defined cut-off frequency. Equalizers on the other hand offer the ability to apply various levels of boost or attenuation at selected frequencies.

2.2.1 Filters: Low-Pass and High-Pass

High-pass and low-pass filters remove frequencies above or below a defined cut-off frequency. Usually the only adjustable parameter is the cut-off frequency, although some models do offer the ability to control the slope of a filter, or how quickly the output drops off beyond the cut-off frequency. Figures 2.2 and 2.3 show frequency response curves for low-pass and high-pass filters, respectively. In practice, high-pass filters are generally employed more often than low-pass filters. High-pass filters can remove low-frequency rumble from a signal, with the engineer taking care to make

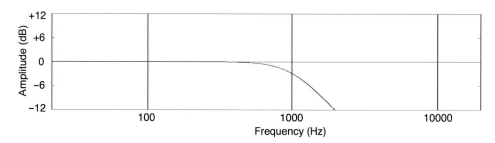

Figure 2.2 The frequency response of a low-pass filter set to 1000 Hz.

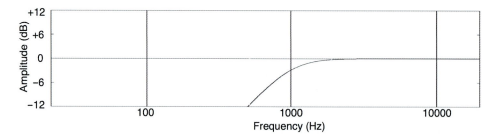

Figure 2.3 The frequency response of a high-pass filter set to 1000 Hz.

sure the cut-off frequency is set below the lowest frequency being produced by the musical instrument signal.

2.2.2 Graphic Equalizers

Graphic equalizers allow control of only the amount of boost or cut for a given set of frequencies, usually with vertical sliders on the front panel of the device. The frequencies available for manipulation are typically based on the International Standards Organization (ISO) center frequencies, such as octave frequencies 31.5 Hz, 63 Hz, 125 Hz, 250 Hz, 500 Hz, 1000 Hz, 2000 Hz, 4000 Hz, 8000 Hz, and 16,000 Hz. It is also possible for a graphic equalizer to have a greater number of bands with increased frequency resolution, such as 1/3rd octave or 1/12th octave frequencies. The bandwidth or Q of each boost or cut is often predetermined by the designer of the equalizer and generally cannot be changed by the user. The graphic equalizer gets

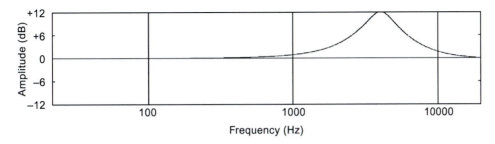

Figure 2.4 The frequency response of a parametric equalizer with a boost of 12 dB at 4000 Hz and a Q of 2.

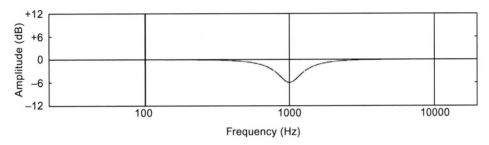

Figure 2.5 The frequency response of a parametric equalizer with a cut of 6 dB at 1000 Hz and a Q of 2.

its name from the fact that the vertical sliders form the shape of the equalization curve from low frequencies on the left to high frequencies on the right.

2.2.3 Parametric Equalizers

A term originally coined by George Massenburg in his 1972 Audio Engineering Society convention paper, the parametric equalizer allows completely independent and sweep-tunable control of three parameters per band: center frequency, Q, and amount of boost or cut at that frequency. The Q is inversely related to the bandwidth of the boost or cut and is defined specifically as follows:

$$Q = F_c/\text{bandwidth}$$

F_c is the center frequency, *bandwidth* is defined as the $f_2 - f_1$. The two frequencies, f_1 and f_2, are the points at which the frequency response is -3 dB down from the maximum boost or $+3$ dB up from the maximum cut.

Figures 2.4 and 2.5 illustrate the frequency responses of two different parametric equalizer settings.

In practice we find that many equalizers are limited in the amount of control they provide. For instance, instead of the Q being fully variable, it may be switchable between three discrete points such as low, mid, high. The selection of center frequency also may not be completely variable, and instead restricted a predetermined set of frequencies. Furthermore, some equalizers do not allow independent control of Q and are designed such that the Q changes according to the amount of gain with minimum boost/cut giving the lowest Q (widest bandwidth) and maximum boost/cut giving the highest Q (narrowest bandwidth).

2.2.4 Shelving Equalizers

Sometimes confused with low-pass and high-pass filters, shelving equalizers can be used to alter a range of frequencies by an equal amount. Whereas high- and low-pass filters can only remove a range of frequencies, shelving equalizers can boost or attenuate by varying degrees a range of frequencies. This range of frequencies extends *downward* from the cut-off frequency for a low-shelf, or it extends *upward* from the cut-off frequency for a high-shelf filter. Probably they are most commonly used as *tone* controls in consumer home or car sound systems. Consumers can alter the spectral balance of their home sound reproduction systems through the use of tone controls and "bass" and "treble" control, which are usually shelving filters with a fixed frequency. High shelving filters apply a given amount of boost or cut equally to all frequencies above the cut-off frequency, whereas low shelving filters apply a given amount of boost or cut equally to all frequencies below the cut-off frequency. In the recording studio, shelving filters are often found as a switchable option in the lowest and highest frequency bands in a parametric equalizer. Some equalizer models also offer high- and low-pass filters in addition to shelving filters.

Below are examples of the frequency response of shelf filters in Figures 2.6 and 2.7.

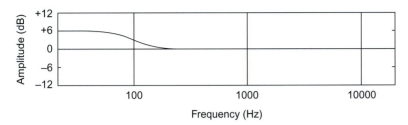

Figure 2.6 The frequency response of a low-shelf filter set to +6 dB 100 Hz.

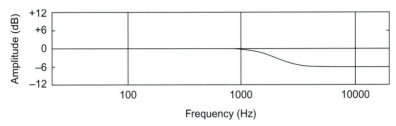

Figure 2.7 The frequency response of a high-shelf filter set to −6 dB at 2000 Hz.

2.3 Getting Started with Practice

It is critical for audio professionals to have a keen sense of spectral balance and how it relates to individual instruments as well as overall mixes. Engineers make decisions about balance of musical elements within an audio recording, and the spectral balance of each individual element within the mix contributes to its ability to blend and "gel" with other elements to form a coherent and clear sonic image. To help develop critical listening skills, a software module is included for the reader to practice hearing the sonic effect of various equalization parameters.

Use of the technical ear training software practice module "TETPracticeEQ" is essential for making progress in accuracy and speed of recognition of equalization. A picture of the user interface is shown in Figure 2.8 and the functionality of the software is described below.

The key to practicing with any of the software modules is maintaining short but regular practice times on a daily or several times a week basis. In the early stages, 10- to 15-minute practice sessions are probably best to avoid getting too

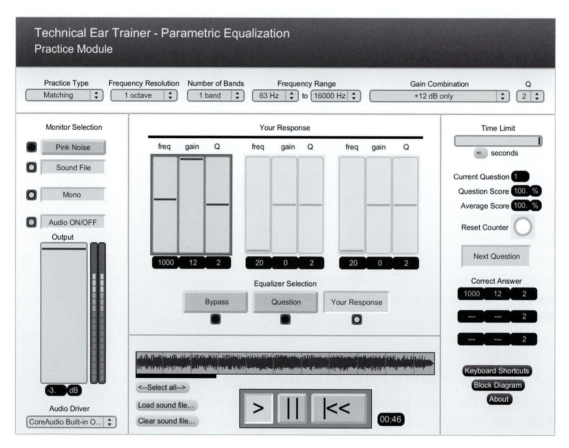

Figure 2.8 A screenshot of the software user interface for the Technical Ear Trainer practice module for parametric equalization.

fatigued. Because of the amount of energy required for highly focused listening, practicing for longer periods of time—a couple of hours or more—typically becomes counterproductive and frustrating. Eventually, as you get used to this focused type of listening, you may want to increase the practice period time but typically 45 to 60 minutes is going to be the upper useful limit for a given practice period. Regular practice for shorter periods of time several times a week is much more productive than extended but less frequent practice sessions. Obviously this could turn into a significant time commitment but taking even 5 minutes a day is likely more effective than trying to cram a 2-hour practice session in once a month.

The software produced for the exercises in this book allows the reader to practice with randomly generated equalization settings within certain limitations chosen by the reader. A screen shot in Figure 2.8 shows the software module for parametric equalization. The goal of the practice module is to identify by ear the equalization parameter settings chosen by the software. The following sections describe the main functions of the software and the available user parameters.

2.3.1 Practice Types

Starting at the top left corner of the window just below the blue header, there is an option to select one of the four practice types: Matching, Matching Memory, Return to Flat, and Absolute Identification:

- *Matching.* Working in Matching mode, the goal is to duplicate the equalization that has been applied by the software. This mode allows free switching between the "Question" and "Your Response" to determine if the chosen equalization matches the unknown equalization applied by the computer.

- *Matching Memory.* This mode is similar to the Matching mode with one main difference—once gain or frequency is changed, the "Question" is no longer available for auditioning. "Question" and "Bypass" are available to be auditioned freely before any changes to the equalizer are made. Matching Memory mode helps us match sounds by memory and can be considered moderately to very difficult depending on the other practice parameters that are chosen, such as number of bands, time limit, and frequency resolution.

- *Return to Flat.* In this mode the goal is to reverse or cancel the randomly chosen equalization applied to the audio signal by the computer by selecting the correct frequency and applying equal but opposite gain to what the software has applied. It is similar in difficulty to "Matching" but requires thinking in the opposite way since the goal is to remove the equalization and return the sound to its original spectral balance. For instance, if you hear a boost of 12 dB at 2000 Hz, the correct response would be to apply a cut of −12 dB at 2000 Hz, thus returning the audio signal to its original state and sounding identical to the "Flat" option. Because the equalization used is reciprocal peak/dip, it is possible to completely eliminate any frequency boosts or cuts by applying equal but opposite boosts or cuts to the respective frequencies. It should be noted that, if you wish to try these exercises in a different context outside of the included software practice modules, not all types of parametric equalizers available are reciprocal peak/dip and thus will not be capable of canceling a boost with an equal but opposite cut. This is not a deficiency but simply a difference in design.

- *Absolute Identification.* This practice mode is the most difficult and the goal is to identify the applied equalization without having the opportunity to listen to what is chosen as the correct response. Only "Bypass" (no equalization) and "Question" (the computer's randomly chosen equalization) can be auditioned.

2.3.2 Frequency Resolution

There are two frequency resolutions from which to choose:
- 1 octave—the easiest of the two options with 9 possible frequencies
- 1/3rd octave—the most difficult with 25 possible frequencies

The frequencies correspond to the International Standards Organization (ISO) frequencies that are common on all commercially available graphic equalizers, as listed in Table 2.1. The software randomly chooses from among these frequencies to apply equalization to the audio signal. Exercises using third-octave frequency resolution are predictably more difficult than those with octave frequencies. The list of third-octave frequencies includes all of the octave frequencies with the addition of two frequencies between each pair of octave frequencies.

It is critical to work with octave frequencies until you excel at identifying all nine octave frequencies. Once these frequencies are solidified, exercises with third-octave frequencies can begin. The octave frequencies should seem like solid *anchors* in the spectrum around which you can identify third-octave frequencies.

One key strategy for identifying third-octave frequencies is to first identify the closest octave frequency. Based on a

Table 2.1 The Complete List of Frequencies (in Hz) shown with Octave Frequencies in Bold

	100	200	400	800	1600	3150	6300	12,500
63	**125**	**250**	**500**	**1000**	**2000**	**4000**	**8000**	**16,000**
80	160	315	630	1250	2500	5000	10,000

solid knowledge of the octave frequencies, you can identify if the frequency in question is in fact one of the nine octave frequencies. If the frequency in question is not an octave frequency, then you can determine if it is *above* or *below* the nearest octave frequency.

For instance, here are two specific octave frequencies (1000 Hz and 2000 Hz) with the respective neighboring third-octave frequencies:

2500 Hz—upper neighbor
2000 Hz—octave frequency anchor
1600 Hz—lower neighbor
1250 Hz—upper neighbor
1000 Hz—octave frequency anchor
800 Hz—lower neighbor

2.3.3 Number of Bands

You can choose to work with one, two, or three frequency bands. This setting refers to the number of simultaneous frequencies that are affected in a given question. The more simultaneous frequency bands chosen, the more difficult a question will be. It is important to work with one frequency band until you are comfortable with the octave and third-octave frequencies. Progressing into two or three bands is much more difficult and can be frustrating without developing confidence with a single band.

When working with more than one band at a time, it can be confusing to know what frequencies have been altered. The best way to work with two or three bands is to identify the most obvious frequency first and then compare your response to the equalizer question. If the frequency chosen does in fact match one of the question frequencies, that particular frequency will become less noticeable when switching between the question and your response, and the remaining frequencies will be easier to identify. The software can accept the frequencies in any order. When working with fewer than three frequency bands, only the leftmost equalizer faders are active.

2.3.4 Frequency Range

We can limit the range of testable frequencies from the full range of 63 Hz to 16,000 Hz to a range as small as three octaves. Users are encouraged to limit the frequency range in the beginning stages to only three frequencies in the midrange, such as from 500 to 2000 Hz. Once these frequencies are mastered, the range can be expanded one octave at a time.

After working up to the full range of frequencies, you may find there remain some frequencies that are still giving you trouble. For instance, low frequencies (in the 63 Hz to 250 Hz range) are often more difficult to identify correctly when practicing with music recordings, especially with third-octave frequencies. This low-frequency range can pose problems because of a number of possible conditions. First, music recordings do not always contain consistent levels across the low-frequency range. Second, the sound reproduction system you are using may not be capable of producing very low frequencies. Third, if it is accurately reproducing low frequencies, room modes (resonant frequencies within a room) may be interfering with what you hear. Using headphones can eliminate any problems caused by room modes, but the headphones may not have a flat frequency response or may be weak in their low-frequency response. For recommendations on specific headphone models, see Section 1.3.3.

2.3.5 Gain Combination

The gain combination option refers to the possible gains (boost or cut) that can be applied to a given frequency. For each question, the software randomly chooses a boost or cut (if there is more than one possible gain) from the gain combination selected and applies it to a randomly selected frequency. When there is only one possible gain, the gain will automatically jump to the appropriate gain when a frequency is chosen.

As one would expect, larger changes in gain (12 dB) are easier to hear than smaller changes in gain (3 dB). Boosts are typically easier to identify than cuts, so it's best to start with boosts until one has become proficient in their identification. It is difficult to identify something that has been removed or reduced but by switching from the equalized version to bypass, it is possible to hear the frequency in question reappear, almost as if it has been boosted above normal.

When working with one band and a gain combination that includes a boost and a cut—such as $+/-6$ dB—it is possible that a low cut can be confused with a high boost and vice versa. A sensitivity to relative changes in frequency response may make a cut in the low frequency range sound like a boost in the high-frequency range.

2.3.6 Q

The Q is a static parameter for any exercise. The default setting of $Q = 2$ is the best starting point for all exercises. Higher Q's (narrower bandwidth) are more difficult to identify.

2.3.7 Sound Source

Practice can be conducted with either pink noise, which is generated internally in the software, or with any two-channel sound file in the format of AIFF or WAV at 44,100- or 48,000-Hz sampling rates. Averaged over time, pink noise has equal power per octave, and its power spectrum appears as a flat line when graphed logarithmically. It also *sounds* equally balanced from low to high frequencies because the auditory system is sensitive to octave relationships (logarithmic) between frequencies rather than linear differences. The range of 20 to 40 Hz represents one octave (a doubling of frequency) but a difference of only 20 Hz, whereas the range between 10,000 Hz and 20,000 Hz is also one octave but a difference of 10,000 Hz. The auditory system perceives both of these ranges as being the same interval: one octave. In pink noise, both of these octave ranges—20 to 40 Hz and 10,000 to 20,000 Hz—have the same power. By using an audio signal that has equal power across the spectrum, we can be assured that a change at one frequency will likely be as audible as a change at any other frequency.

There is also an option to listen to the sound source in mono or stereo. If a sound file loaded in contains only one track of audio (as opposed to two), the audio signal will be sent out of the left output only. By pressing the mono button, the audio will be fed to both left and right output channels.

It is best to start with pink noise when beginning any new exercises and subsequently practice with recordings of various instrumentation and genres. The greater the variety of sound recordings used, the more able you will be able to transfer the skills obtained in these exercises to other listening situations.

2.3.8 Equalizer Selection

In the practice software, an audio signal (pink noise or audio file signal) is routed to three places:
- Straight through with no equalization—bypassed
- Through the "Question" equalizer chosen by the computer
- Through the user equalizer ("Your Response")

We can select which of these options to audition. The Bypass selection allows us to audition the original audio signal without any equalization applied. The selection labeled "Question" allows us to audition the equalization that has been randomly chosen by the software and applied to the audio signal. The selection labeled "Your Response" is the equalization applied by the user, according to the parameters shown in the user interface. See Figure 2.9, which shows a block diagram of the practice module.

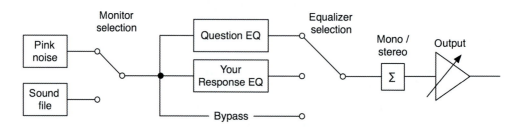

Figure 2.9 A block diagram of the signal path for the Technical Ear Trainer practice module for parametric equalization.

2.3.9 Sound File Control

The Sound File Control section of the interface includes a waveform display of the audio signal. You can select excerpts of the full audio file by clicking and dragging on the waveform. The audio file automatically repeats once it reaches the end of the file or the end of the selected section. By simply clicking in the waveform, the waveform is selected from the location of the click to the end of the file.

2.3.10 Time Limit

In the recording studio or live sound venue, time is of the essence. Engineers must often make quick and accurate decisions about sound quality and audio signal processing. To help prepare for these real-world situations, a time limit can be applied in the practice module so that you can practice identifying equalization parameters with speed as well as accuracy.

The keyboard shortcuts included in the software are ideal for quickly indicating responses when using the timer. When working on exercises with more than one frequency band, the tab key cycles through bands. The up/down arrows can be used to increase or decrease octave frequencies. Alternatively, number keys correspond to octave frequencies (0 = 20 Hz, 1 = 63 Hz, 2 = 125 Hz, 3 = 250 Hz, 4 = 500 Hz, 5 = 1000 Hz, 6 = 2000 Hz, 7 = 4000 Hz, 8 = 8000 Hz, and 9 = 16,000 Hz) and can be used to jump to an octave frequency immediately. The left/right arrows adjust the gain of a selected band in 3-dB increments. For exercises with only one gain option (e.g., +12 dB), the gain is automatically set when the frequency slider is changed from 20 Hz to any other frequency. Returning the frequency slider to 20 Hz resets the gain to 0 dB. For exercises with more than one gain option (e.g., +/−12 dB), the gain stays at 0 dB until adjusted by the user; it does not automatically change when the frequency is changed.

Sometimes a time limit is useful in that it forces us to respond with our first impression rather than spend too much time thinking and rethinking. Novice recording engineers who have spent time with the practice module have often reported that overthinking a question results in mistakes and that their first impressions are often the most accurate.

2.3.11 Keyboard Shortcuts

- [space bar] toggles the Equalizer Selection depending on the Practice Type:
  - Matching: toggles between Question and Your Response
  - Matching Memory: toggles between Question and Your Response, until a parameter is changed at which point it toggles between Bypass and Your Response
  - Return to Flat: toggles between Your Response and Bypass
  - Absolute Identification: toggles between Question and Bypass
- [enter] or [return] checks answer and moves to next question
- [q] listen to Bypass
- [w] listen to Question
- [e] listen to Your Response
- Numbers 1 to 9 correspond to octave frequencies of a selected band (e.g., 1 = 63 Hz, 2 = 125 Hz, 3 = 250 Hz, 4 = 500 Hz, 5 = 1000 Hz, 6 = 2000 Hz, 7 = 4000 Hz, 8 = 8000 Hz, 9 = 16,000 Hz)
- Up/down arrows change the frequency of the selected band
- Left/right arrows change the gain of the selected band
- [tab] selects the frequency band to modify, if number of bands is more than one
- [esc] turns audio off

2.4 Working with the EQ Practice Module

Upon first opening up the EQ practice module, select pink noise in the Monitor Selection, turn the audio on, and adjust the output level to a comfortable listening level. Make sure the Equalizer Selection is set to Your Response, and scroll through each octave frequency to get a feel for the sound of each frequency. Once you change the frequency, the gain will automatically jump to 12 dB; this is the default gain combination setting upon opening the software module. Switch between Bypass (no equalization) and Your Response to compare the change in timbre that is created by a boost at each frequency. Spend some time initially just listening to various frequencies, alternating between flat and equalized. After familiarizing yourself with how the octave frequencies sound with pink noise, load in a sound file and do the same thing again, auditioning all octave frequencies.

When you are auditioning a sound file, start to make note of what instruments or components of instrument sounds are affected by each particular octave frequency. For instance, 125 Hz may bring out the low harmonics in a snare drum or bass. On the upper end of the spectrum, 8 kHz may bring out crisp cymbal harmonics. If you are auditioning a Baroque ensemble recording, you may find that a boost at 8 kHz makes a harpsichord more prominent. Boosts at specific frequencies can sometimes bring out individual instruments in a mix and in fact skilled mastering engineers use this capability to provide subtle rebalancing of a mix.

Each recording will be affected slightly differently by a given frequency, even with comparable instrumentation. Depending on the frequency content and spectral balance of each individual instrument in a recording, the effect of an equalizer setting will be somewhat different from mix to mix. This is one reason why an engineer must be attentive to what is required in each individual recording, as opposed to simply relying on what may have worked in previous recordings. For example, just because a cut at 250 Hz may have worked on one snare drum in one recording does not mean that it will work on all snare drum recordings.

Sometimes during the recording and mixing process, we may find ourselves evaluating and questioning our processing and mixing decisions based on logic of what seems correct from a numerical point of view. For instance, let's say we apply a cut of 20 dB at 300 Hz on an individual instrument. There can be a temptation to evaluate the amount of equalization and think that 20 dB is too much, based on what would *seem* reasonable (i.e., thinking to ourselves, "I have never had to do this before and it seems like an extreme setting, so how can it be right?") rather than what *sounds* reasonable. The evaluation of a decision based on what we *think* is appropriate does not always coincide with what clearly *sounds* most appropriate. In the end, it does not matter how ridiculous a signal processing or mixing decision may appear as long as the sonic result suits the artistic vision we have for a project. As an engineer, we can have a direct effect on the artistic impression created by recorded music depending on choices such as balance and mix levels, timbre, dynamics, and spatial processing. Judgments about what is appropriate and suitable should be made by ear with no judgment of the actual parameter numbers being chosen.

2.4.1 Vowel Sounds

A number of researchers have noted that associating specific vowel sounds with octave frequencies can help listeners identify frequencies because of the formant frequencies present in each vowel sound (Letowski, 1985; Miskiewicz, 1992; Opolko & Woszczyk, 1982; Quesnel, 2001; Quesnel & Woszczyk, 1994; Slawson, 1968). The following vowel sounds roughly correspond to octave frequencies:

- 250 Hz = [u] as in b<u>oo</u>t
- 500 Hz = [o] as in t<u>ow</u>
- 1000 Hz = [a] as in f<u>a</u>ther
- 2000 Hz = [e] as in b<u>e</u>t
- 4000 Hz = [i] as in b<u>ee</u>t

Matching frequency resonances to specific vowel sounds can help with learning and memory of these particular frequencies. Instead of trying to think of a frequency number, some readers will find it useful to match the sound they are hearing with a vowel sound. The vowel sound can then be linked to a specific octave frequency.

2.5 Recommended Recordings for Practice

The following list identifies some commercially available recordings of various genres that are suitable for use as sound sources in the EQ software practice module. They represent examples of high-quality recordings that have good spectral balance across a wide frequency range. Compact Disc quality versions should be used (i.e., digital linear pulse-code modulation 44.1 kHz, 16-bit AIFF or WAV) for all exercises. Encoded versions (such as MP3, Windows Media Audio, or Advanced Audio Coding) should never be used for EQ exercises, even if they have been converted back to PCM. Once an audio file has been perceptually encoded, its quality has been degraded and cannot be recovered by converting back to linear PCM.

Anderson, Arild. (2004). "Straight" from *The Triangle*. ECM Records. (jazz piano trio)

Blanchard, Terence. (2001). "On the Sunny Side of the Street" from *Let's Get Lost*. Sony. (jazz with vocals)

Earth, Wind & Fire. (1998). "September" from *Greatest Hits*. Sony. (R&B pop)

Hellendaal, Pieter. (1991). "Concerto II—Presto" from *6 Concerti Grossi*. Perf. The European Community Baroque Orchestra. Channel Classics. (Baroque orchestra)

Le Concert des Nations. (2002). "Marche pour la cérémonie" from *Soundtrack from the film Tous les matins du monde*. Alia Vox Spain. (Baroque orchestra)

Randall, Jon. (2005). *Walking Among the Living*. Epic/Sony BMG Music Entertainment. (roots music/bluegrass)

Steely Dan. (2000). "Gaslighting Abbie" from *Two Against Nature*. Giant Records. (pop)

The Police. (1983). "Every Breath You Take" from *Synchronicity*. A&M Records. (rock)

There are also a few artists that are making multitrack stems available for purchase or free download. Apple's GarageBand and Logic also offer recordings of solo instruments that can be useful with the software.

Summary

Equalization is one of the most important tools of any audio engineer. It is possible to learn how to identify resonances and antiresonances by ear through practice. The included software practice module can serve as an effective tool for progress in technical ear training and critical listening when used for regular and consistent practice.

3

SPATIAL ATTRIBUTES AND REVERBERATION

Audio Production and Critical Listening. DOI: 10.1016/B978-0-240-81295-3.00003-4

Reverberation is used to create distance, depth, and spaciousness in recordings, whether captured with microphones during the recording process or added later during mixing. In classical music recording, engineers strive to achieve a fairly natural representation of a musical ensemble on a stage in a reverberant performance space. In this type of recording, microphones are positioned to capture direct sound arriving straight from the instruments as well as indirect sound reflected back from a surrounding enclosure (walls, ceiling, floor, seats). Engineers seek to achieve an appropriate balance of direct and indirect sound by adjusting the locations and angles of microphones.

Pop, rock, electronic, and other styles of music that use predominantly electric instruments, and computer-generated sounds are not necessarily recorded in reverberant acoustic spaces. Rather, a sense of space present is often created through the use of artificial reverberation and delays, after the music has been recorded in a relatively dry acoustic space. Artificial reverberation and delay are used both to mimic real acoustic spaces and to create completely unnatural sounding spaces.

Delay and reverberation help create a sense of depth and distance in a recording, helping to position some sound sources farther away (i.e., upstaging them) while other less reverberant elements remain to the front of a phantom image sound stage. Not only can an engineer make sounds seem farther away and create the impression of an acoustic space, but he can influence the character and mood of a music recording with careful use of reverberation. In addition to depth and distance control, the angular location of sound sources is controlled through amplitude panning. When listening over speakers, an engineer has essentially two dimensions within which to control a sound source location: distance and angular location (azimuth).

Taken as a whole, we can consider the properties of sound source location within a simulated acoustic space, the qualities of a simulated acoustic space, as well as the coherence and spatial continuity of a sound image collectively as the spatial attributes of a recording.

3.1 Analysis of Perceived Spatial Attributes

The auditory system extracts information about the spatial attributes of a sound source, whether the source is an acoustic musical instrument or a recording of a musical instrument reproduced over loudspeakers. Spatial attributes help determine with varying levels of accuracy the azimuth, elevation, and distance of sound sources, as well as information about the environment or enclosure in which they are produced. The binaural auditory system relies on interaural time differences, interaural intensity differences, and filtering by the pinnae or outer ear to determine the location of a sound source (Moore, 1997). The process of localization of sound images reproduced over loudspeakers is somewhat different from localization of single acoustic sources, and in this chapter we will concentrate on the spatial attributes that are relevant to audio production and therefore sound reproduction over loudspeakers.

Spatial attributes include the perceived layout of sources in a sound image, characteristics of the acoustic environment in which they are placed, as well as the overall quality of a sound image produced by loudspeakers. It is critical for a recording engineer to have a highly developed sense for any spatial processing already present in or added to a recording. Panning and spatial effects have a great effect on the balance and blend of elements in a mix, which in turn influence the way in which listeners perceive a musical recording. For example, the use of a longer reverberation time can create drama and excitement in a music recording by creating the impression that the music is emanating from a big space. Alternatively, with the use of short reverberation times, an engineer can create a sense of intimacy or starkness to the music.

The spatial layout of sources in a sound image can influence clarity and cohesion in a recording as spatial masking plays a role in the perceived result. Occasionally the use of reverberation in a sonically dense recording can seem inaudible or at least difficult to identify because it blends with and is partially masked by direct sound. When mixing a track with a small amount of reverb, there are times when it is helpful to mute and unmute any added reverberation to hear its contribution to a mix.

As we consider the parameters available on artificial reverberation such as decay time, predelay time, and early reflections, we must also factor in subjective impressions of spatial processing as we translate between controllable parameters and their sonic results. For example, there is usually not a parameter labeled "distance" in a reverberation processor, so if we want to make a sound source more distant, we need to control distance indirectly by adjusting parameters in a coordinated way until we have the desired sense of distance. An engineer must translate between objective parameters of reverberation to create the desired subjective impression of source placement and simulated acoustic environment. It is difficult to separate sound source distance control from the simulation of an acoustic environment, because an integral part of distance control is the creation of a perceived sound stage within a mix—a virtual environment from which musical sounds appear to emanate.

The choice of reverberation parameter settings depends on a number of things such as the transient nature and width of a dry sound source, as well as the decay and early reflection characteristics of a reverberation algorithm. Professional engineers often identify subjective qualities of each reverb that bring them closer to their specific goals for each mix rather than simply choosing parameter settings that worked in other situations. A particular combination of parameter settings for one source and reverberation usually cannot simply be duplicated for an identical distance and spaciousness effect with a different source or reverberation.

We can benefit from analyzing spatial properties from both objective and subjective perspectives, because the tools have objective parameters, but our end goal in recording is to achieve a great sounding mix, not to identify specific parameter settings. As with equalization, we must find ways to translate between what we hear and the parameters available for control. Spatial attributes can be broken down into the following categories and subcategories:

- Placement of direct/dry sound sources
- Characteristics of acoustic spaces and phantom image sound stages
- Characteristics of an overall sonic image produced by loudspeakers

3.1.1 Sound Sources

3.1.1.1 Angular Location

Also called azimuth, the angular location of a sound source is its perceived location along the horizontal plane relative to the left and right loudspeakers. Typically it is best to spread out sources across the stereo image so that there is less masking and more clarity for each sound source. Sounds can mask one another when they occupy a similar frequency range and angular location.

Each microphone signal can be panned to a specific location between loudspeakers using conventional constant-power panning found on most mixers. Panning can also be accomplished by delaying a signal's output to one loudspeaker channel relative to the other loudspeaker output. Using delay for panning is not common because its effectiveness depends highly on a listener's location relative to the loudspeakers.

Balancing signals from a few a stereo microphone techniques will usually require panning of each pair of microphone signals hard left and hard right. The resulting positions of sound sources that are in front of each microphone pair will depend on the stereo microphone technique used and the respective locations of each source.

3.1.1.2 Distance

Although human perception of absolute distance is limited, relative distance of sounds within a stereo image is important to give depth to a recording. Large ensembles recorded in acoustically live spaces are likely to exhibit a natural sense of depth, analogous to what we would hear as an audience member in the same space. With recordings made in acoustically dry spaces such as studios, engineers often seek to create depth using delays and artificial reverberation. Engineers can control sound source distance by adjusting physical parameters such as the following:

- *Direct sound level.* Quieter sounds are judged as being farther away because there is a sound intensity loss of 6 dB per doubling of distance from a source. This cue can be ambiguous for the listener because a change in loudness can be the result of either a change in distance or a change in a source's acoustic power.

- *Reverberation level.* As a source moves farther away from a listener in a room or hall, the direct sound level decreases and the reverberant sound remains the same, lowering the direct-to-reverberant sound ratio.
- *Distance of microphones from sound sources.* Moving microphones farther away decreases the direct-to-reverberant ratio and therefore creates a greater sense of distance.
- *Room microphone placement and level.* Microphones placed at the opposite side of a room or hall from where musicians are located pick up sound that is primarily reverberant or diffuse. Room microphone signals can be considered as a reverberation return on a mixer.
- *Low-pass filtering of close-miked direct sounds.* High frequencies are attenuated more than lower frequencies because of air absorption. Furthermore, the acoustic properties of reflective surfaces in a room affect the spectrum of reflected sound reaching a listener's ears.

3.1.1.3 Spatial Extent

Sometimes sound source locations in a mix are precisely defined, where other times sound source location is fuzzier and more difficult to pinpoint. Spatial extent describes a source's perceived width. A related concept in concert hall acoustics research is apparent source width (ASW), which is related to strength, timing, and direction of lateral reflections. Barron (1971) found that stronger lateral reflections would result in a wider ASW.

The perceived width of a sound image produced over loudspeakers will vary with the microphone technique used and the sound source being recorded. Spaced microphones produce a wider sound source because the level of correlation between the two microphone signals is reduced as the microphones are spread farther apart. As with concert hall acoustics, perceived width of sources reproduced over loudspeakers can also be influenced by early reflections, whether recorded with microphones or generated artificially. If artificial early reflections (in stereo) are added to a single, close microphone recording of a sound source, the direct sound tends to fuse perceptually with early reflections and produce an image that is wider than just the dry sound on its own.

Spatial extent of sound sources can be controlled through physical parameters such as the following:

- Early reflection patterns originating from a real acoustic space or generated artificially with reverberation
- Type of stereo microphone technique used: spaced microphones generally yield a wider spatial image than coincident microphone techniques

3.1.2 Acoustic Spaces and Sound Stages

An engineer can control additional spatial attributes such as the perceived characteristics, qualities, and size of the acoustic environment in which each sound source is placed in a stereo image. The environment or sound stage may consist of a real acoustic space captured with room microphones, or it may be created by artificial reverberation added during mixing. There may be a common type of reverberation for all sounds, or some sounds may have unique types of reverberation added to them to help set them apart from the rest of the instruments. For instance, it is fairly common to treat vocals or solo instruments with a different reverberation than the rest of an accompanying ensemble.

3.1.2.1 Reverberation Decay Character

Decay time is one of the most commonly found parameters on artificial reverberation devices. In recording acoustic instruments in a live acoustic space, the reverberation decay time is often not adjustable, yet some recording spaces have been designed with panels on wall and ceiling surfaces that can be rotated to expose various sound absorbing or reflecting materials, allowing a somewhat variable reverberation decay time.

The decay time is defined as the time in which sound continues to linger after the direct sound has stopped sounding. Longer reverberation times are typically more audible than shorter reverberation times for a given reverberation level. Transient sounds such as drums or percussion expose decay time more than sustained sounds, allowing us to hear the rate of decay more clearly.

Some artificial reverberation algorithms will incorporate modulation into the decay to give it variation and hopefully make it sound less artificial. A perfectly smooth decay is something that we rarely hear in a real room, and a simplified artificial reverberation can sound unnaturally smooth.

3.1.2.2 Spatial Extent (Width and Depth) of the Sound Stage

A sound stage is the acoustic environment within which a sound source is heard, and it should be differentiated from a sound source. The environment may be a recording of a real space, or it may be something that has been created artificially using artificial delay and reverberation.

3.1.2.3 Spaciousness

Spaciousness represents the perception of physical and acoustical characteristics of a recording space. In concert hall acoustics, it is related to envelopment, but with only two loudspeakers in stereo reproduction, it is difficult to achieve true envelopment. We can use the term *spaciousness* to describe the feeling of space within a recording.

3.1.3 Overall Characteristics of Stereo Images

Also grouped under spatial attributes are items describing overall impressions and characteristics of a stereo image reproduced by loudspeakers. A stereo image is the illusion of sound source localization from loudspeakers. Although there are only two loudspeakers for stereo, the human binaural auditory system allows for the creation of phantom images at locations between the loudspeakers. In this section, we consider the overall qualities of a stereo image that are more generalized than those specific to the source and sound stage.

3.1.3.1 Coherence and Relative Polarity between Channels

Despite widespread use of stereo and multichannel playback systems among consumers, mono compatibility continues to remain critically important, mainly because we can to listen to music through computers and mobile phones with single speakers. Checking a mix for mono compatibility involves listening for changes in timbre that result from destructive interference between the left and right channels. In the worst-case scenario with opposite polarity stereo channels, summation to mono will cancel a significant portion of a mix. Each project that an engineer mixes needs to be checked to make sure that the two channels of a stereo mix are not opposite polarity. When left and right channels are both identical and opposite polarity, they will cancel completely when summed together. If both channels are identical, then the mix is monophonic and not truly stereo. Most stereo mixes include some combination of mono and stereo components. We can describe the relationship between signal components in the left and right channels of a mix as existing along a scale of correlation between -1 and 1:

- Left and right are identical—composed of signals that are panned center, having a correlation of 1
- Left and right share nothing in common—signals that are panned to one side or the other, or similar signals having a correlation of 0 between channels
- Left and right channels are identical but opposite in polarity—signals having a correlation of -1

Phase meters provide one objective way of determining the relative polarity of stereo channels, but if no such meters are available, an engineer must rely on her ears. Opposite polarity left and right channels can be identified by listening for an extremely wide stereo image, such that when sitting in the ideal listening position (see Fig. 1.2), sound from the loudspeakers seems to come from the sides. Another characteristic of opposite polarity channels is that the stereo image is unstable and tends to move in an exaggerated way with small head movements. Section 3.7.3 offers more information on auditioning opposite polarity channels.

On occasion an individual instrument may be represented in a mix by two identical but opposite polarity signals panned hard right and left. If such a signal is present, a phase meter may not register it strongly enough to give an unambiguous visual indication. Sometimes stereo line outputs from electric instruments are opposite polarity or perhaps a polarity flip cable was used during recording by mistake. Often stereo (left and right) outputs from electronic instruments are not truly stereo but mono. When one output is of opposite polarity, the two channels will cancel when summed to mono.

3.1.3.2 Spatial Continuity of a Sound Image from One Loudspeaker to Another

As an overall attribute, mixing engineers consider the continuity and balance of a sound image from one loudspeaker to another. An ideal stereo image will be balanced between left and right and will not have too much or too little energy located in the center. Often pop and rock music mixes have a strong center component because of the number and strength of instruments that are typically panned center, such as kick drum, snare drum, bass, and vocals. Classical and acoustic music recordings may not have a similarly strong central image, and it is possible to have a deficiency in the amount of energy in the center—sometimes referred to as having a "hole in the middle." Engineers strive to have an even and continuous spread of sound energy from left to right.

3.2 Basic Building Blocks of Digital Reverberation

Next we will explore two fundamental processes found in most digital reverberation units: time delay and reverberation.

3.2.1 Time Delay

Although a simple concept, time delay can serve as a fundamental building block for a wide variety of complex effects. Figure 3.1 shows a block diagram of a single delay combined with a nondelayed signal. Figure 3.2 shows what the output of the block diagram would look like if the input was an impulse.

By simply delaying an audio signal and mixing it with the original nondelayed signal, the product is either comb filtering (for shorter delay times) or echo (for longer delay times). By adding hundreds of delayed versions of a signal in an organized way, early reflection patterns such as those found in real acoustic spaces can be mimicked. Chorus and flange effects are created through the use of delays that vary over time.

3.2.2 Reverberation

Whether originating from a real acoustic space or an artificially generated one, reverberation is a powerful effect that

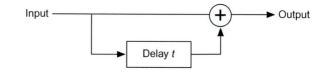

Figure 3.1 A block diagram of a delay line.

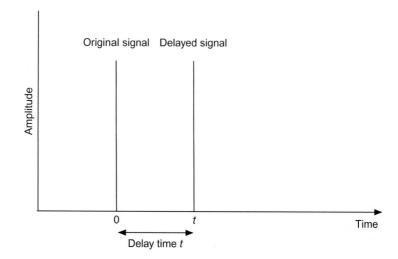

Figure 3.2 A time-based view of the output of a signal (in this case an impulse) plus a delayed version of itself.

provides a sense of spaciousness, depth, cohesion, and distance in recordings. Reverberation helps blend recorded tracks to create a unified sound image where all the components in an image reside in a common acoustic space. In reproduced sound, reverberation can create the illusion of being immersed in an environment that is different from our physical surroundings.

On the other hand, reverberation, like any other type of audio processing, can also create problems in sound recording and production. Reverberation that is too high in level or too long in decay time can destroy the clarity of direct sounds or, as in the case of speech, affect the intelligibility of what is being said. The quality of reverberation must be optimized to suit the musical and artistic style being recorded.

Reverberation and delay have important functions in music recording, such as helping the instruments and voices in a recording blend and "gel." Through the use of reverberation, an engineer can influence a listener's sense of a mix by creating the illusion of sources performing in a common acoustic space. Additional layers of reverberation and delay can be added to accentuate and highlight specific soloists.

The sound of a close-miked instrument or singer played back over loudspeakers creates an intimate or perhaps even uncomfortable feeling for a listener. Hearing such a recording over headphones can create the impression that a singer is only a few centimeters from the ear, and this is not something listeners are accustomed to hearing acoustically from a live music performance. Live music performances are typically heard at some distance away, which means that reflected sound from walls, floor, and ceiling of a room fuses perceptually with sound coming directly from a sound source. When using close microphone placement in front of a musical performer, it often helps to add some delay or reverberation to the "dry" signal to create some perceived distance between the listener and sound source.

Conventional digital reverberation algorithms use a network of delays, all-pass filters, and comb filters as their building blocks, based on Schroeder's (1962) original idea (Fig. 3.3). Equalization is applied to alter the spectral content of reflections and reverberation. In its simplest form, artificial reverberation is simply a combination of delays with feedback or recursion. Each time a signal goes through the feedback loop it is reduced in level by a preset amount so that its strength decays over time.

More recent reverberation algorithms have been designed to convolve an impulse response of a real acoustic space with the incoming "dry" signal. Hardware units capable of convolution-based reverberation have been commercially available since the late 1990s, and software implementations are now commonly released as plug-ins with digital audio workstation software. Convolution reverberation is sometimes referred to as sampling reverb because a "sample" of an acoustic space (i.e., its impulse response) is convolved with a dry audio signal. Although it is possible to compute in the time domain, the convolution process is usually completed in the frequency domain to make the computation fast enough for real-time processing. The resulting audio signal from a convolution reverberator is arguably a more realistic sounding reverberation than what is possible with conventional digital reverberation. The main drawback is that there is not as much flexibility or control of parameters of the convolution

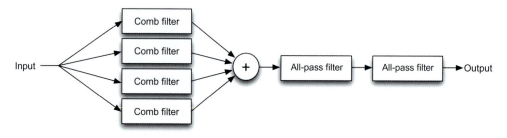

Figure 3.3 A block diagram of Manfred Schroeder's original digital reverberation algorithm.

reverberation as is possible with digital reverberation based on comb and all-pass filters.

In conventional digital reverberation units, a number of possible parameters are available that can be controlled. Although these parameters vary from one manufacturer to another, a few of the most common include the following:

- Reverberation decay time (RT60)
- Delay time
- Predelay time
- Some control over early reflection patterns, either by choosing between predefined sets of early reflections or control over individual reflections
- Low-pass filter cut-off frequency
- High-pass filter cut-off frequency
- Decay time multipliers for different frequency bands
- Gate—threshold, attack time, hold time, release or decay time, depth

Although most digital reverberation algorithms represent simplified models of the acoustics of a real space, they are widely used in recorded sound to help augment the recorded acoustic space or to create a sense of spaciousness that did not exist in the original recording environment.

3.2.2.1 Reverberation Decay Time

The reverberation time is defined as the amount of time it takes for a sound to decay 60 dB once it is turned off. Usually referred to as RT60, W.C. Sabine proposed an equation for calculating it in a real acoustic space (Howard & Angus, 2006):

$$RT60_{\alpha<0.3} = 0.161V/S\alpha$$

V = volume in m^3, S = surface area in m^2 for a given type of surface material, and α = absorption coefficient of the respective surface.

Because the RT60 will be some value greater than zero even if α is 1.0 (100% absorption on all surfaces), the Sabine equation is typically only valid for α values less than 0.3. In other words, the shortcoming of the Sabine equation is that even in an anechoic chamber, a reverberation time will still be calculated, even though no reverberation would be measured acoustically. Norris-Eyring proposed a slight variation on the equation for a wider range of values (Howard & Angus, 2006):

$$RT60 = -0.161V/S \ln(1 - \alpha)$$

It is critical for an engineer to have an intuitive sense for what decay times of various values mean in terms of how they sound. A decay time of 2 seconds will have a much different sonic effect in a mix from a decay time of 1 second.

3.2.2.2 Delay Time

A straight delay without feedback or recursion of an audio signal is often mixed with the dry signal to create a sense of space, and it can supplement or be a substitute for the use of reverberation.

With shorter delay times—less than about 30 milliseconds—the auditory system tends to fuse the direct and delayed sounds, judging the position of the combined sound based on the location of the direct sound. The phenomenon is known as the *precedence effect*, the *Haas effect*, or the *law of the first wavefront*. With delay times of more than about 30 milliseconds, the delayed signal is heard as a distinct echo of a direct sound. The actual amount of delay time required to create a distinct echo depends on the nature of the audio signal being delayed. Transient, percussive signals reveal distinct echoes with much shorter delay times (less than 30 milliseconds), whereas sustained, steady-state signals require much longer delay times (more than 50 milliseconds) to create an audible echo.

3.2.2.3 Predelay Time

Predelay time is typically defined as the time delay between the direct sound and the onset of reverberation. Perceptually it can give the impression of a larger space as the predelay time is increased. In a real acoustic space with no physical obstructions between a sound source and a listener, there will always be a short delay between the arrival of direct and reflected sounds. The longer this initial delay is, the larger a space is perceived to be.

3.2.3 Digital Reverberation Presets

Most digital reverberation units currently available, whether in plug-in or hardware form, offer hundreds if not thousands of reverberation presets. What may not be immediately obvious to the novice engineer is that there are typically only a handful of different algorithms for a given type or model of reverberation. The presets are simply the same algorithms repeated with variations on the parameter settings and individually named so as to reflect the type of space the unit is modeling or a possible application such as large hall, bright vocal, studio drums, or theater. All of the presets using a given type of algorithm represent identical types of processes and will sound identical if the parameters of each preset are matched.

Because engineers adjust many reverberation parameters to create the most suitable reverberation for each application, it may make sense to pick any preset and start tuning the parameters as opposed to trying to find a preset that will work without any adjustments. The main drawback of trying to find the right preset for each instrument and voice during a mix is that the "right" preset might not exist and it will likely require adjustment of parameters anyway. It may be best to start right away by choosing any preset and editing the parameters to suit a mix. The process of editing parameters as opposed to trying to find a preset will also help in learning the capabilities of each reverb and the sonic result of each parameter change.

Although it may not be the best use of time to search for a preset during the mixing process, there is an advantage to going through presets and listening to each one because it can give a clearer idea of what a reverberation unit can sound like across many different parameter settings. Such a listening exercise should be done at a time outside of a mix project to allow time for listening and getting familiar with the hardware and software at our disposal.

3.3 Reverberation in Multichannel Audio

From a practical point of view, my informal research and listening seem to indicate that, in general, higher levels of reverberation are possible in multichannel audio recordings than two-channel stereo, while maintaining an acceptable level of clarity. More formal tests need to be conducted to verify this point, but it may make sense from what we know about masking. Masking of one sound by another is reduced when the two sounds are separated spatially (Kidd et al., 1998; Saberi et al., 1991). It appears that because of the larger spatial distribution of sound in multichannel audio, relative to two-channel stereo, reverberation is less likely to obscure or mask the direct sound and therefore can be more prominent in multichannel audio.

One could argue that reverberation is increasingly critical in recordings mixed for multichannel audio reproduction because multichannel audio offers a much greater possibility to re-create a sense of immersion in a virtual acoustic space than two-channel stereo. There has been much more investigation of the spatial dimension of reproduced sound in recent years as multichannel audio has grown in popularity and its distribution has grown to a wider audience. As such, students of recording engineering can benefit from a systematic training method to learn to match parameter settings of artificial reverberation "by ear" and to further develop the ability to consistently identify subtle details of sound reproduced over loudspeakers.

Recording music and sound for multichannel reproduction also presents new challenges over two-channel stereo in terms of creating a detailed and enveloping sound image. One of the difficulties with multichannel audio reproduction using the ITU-R BS.775 (ITU-R, 1994) loudspeaker layout is the large space between the front and rear loudspeakers (80 to 90° spacing; see Fig. 1.3). Because of the spacing between the loudspeakers and the nature of our binaural sound localization abilities, lateral phantom images are typically unstable. Furthermore, it is a challenge to produce phantom images that join the front sound image to the rear. Reverberation can be helpful in creating the illusion of sound images that span the space between loudspeakers.

3.4 Software Training Module

The included software training module is a tool to assist in hearing subtle details and parameters of artificial digital reverberation rather than an ear trainer for the perception of room acoustics. It may be possible that skills obtained from using this system will help in the perception of acoustical characteristics, but it is not clear how well one skill transfers to the other. Most conventional digital reverberation algorithms are based on various combinations of comb and all-pass filters after the model developed by Schroeder, and although these algorithms are computationally efficient and provide many controllable parameters, they are not physical models of the behavior of sound in a real room. Therefore, it is not possible to confirm that parameters of artificial reverberation such as decay time are identical to those found in sound in a real acoustic space. It is not clear how closely the reverberation decay time (RT60) of a given artificial reverberation algorithm relates to decay time of sound in a real room. For instance, if the decay times of different artificial reverb units or plug-ins are set to 1.5 seconds, the perceived decay time may be found to differ between units. Furthermore, reverberation time is sometimes dependent on other parameters in an algorithm. It is not always clear exactly what other parameters such as "size" are controlling or why they might affect the perceived decay time without changing the displayed decay time. Because of the variability of perceived decay time between units and algorithms, it is perhaps best not to learn absolute decay times but rather to learn to hear differences between representative examples and be able to match parameter settings. Nonetheless, reverberation is a powerful sonic tool available to recording engineers who mix it with recorded sound to create the aural illusion of real acoustics and spatial context.

Just as it is critical to train audio engineers to recognize spectral resonances, it is equally important to improve our perception of subtleties in artificial reverberation. At least one researcher has demonstrated that listeners can "learn" reverberation for a given room (Shinn-Cunningham, 2000). Other work in training listeners to identify spatial attributes of sound has been conducted as well. Neher et al. (2003) have documented a method of training listeners to identify spatial attributes using verbal descriptors for the purpose of spatial audio quality evaluation.

Research has been conducted to describe the spatial attributes of reproduced sound using graphical assessment (such as Ford et al., 2003, and Usher & Woszczyk, 2003). An advantage of the training system discussed herein is that you compare one spatial scene with another, by ear, and is never required to translate or map an auditory sensation to a second sensory modality and subsequently to a means of expression, such as drawing an image or choosing a word. Using the system, you can compare and match two sound scenes, within a given set of artificial reverberation parameters, using only the auditory system. Thus, there is no isomorphism between different senses and methods of communication. Additionally this method has ecological validity, as it mimics the process of a sound engineer sculpting sonic details of a sound recording by ear rather than through graphs and words.

3.5 Description of the Software Training Module

The included software training module "TETpracticeReverb" is available for listening drills. The computer randomizes the exercises and gives a choice of difficulty and choice of parameters for an exercise. It works in much the same way as the EQ module described in Chapter 2.

3.5.1 Sound Sources

Readers are encouraged to begin the training course with simple, transient, or impulsive sounds such as percussion and progress to more complex sounds such as speech and music recordings. In the same way that pink noise is used in the initial stages of frequency ear training because it exposes a given amount of equalization better than most music samples, percussive or impulsive sounds are used for beginning levels of training in time-based effects processing because the sonic character of = reverberation is more apparent than with steady-state sources. The temporal character of a sound affects the ability to hear qualities of the reverberation when the two are mixed. Typically transient or percussive sounds reveal reverberation, whereas more steady-state, sustained musical passages tend to mask or blend with reverberation, making judgments about it more difficult.

3.5.2 User Interface

A graphical user interface (GUI), shown in Figure 3.4, provides a control surface for you to interact with the system. With the GUI you can do the following:

- Choose the level of difficulty
- Select the parameter(s) with which to work
- Choose a sound file
- Adjust parameters of the reverberation
- Toggle between the reference and your answer
- Control the overall level of the sound output
- Submit a response to each question and move to the next example

The graphical interface also keeps track of the current question and the average score up to that point, and it provides the score and correct answer for the current question.

3.6 Getting Started with Practice

The training curriculum covers a few of the most commonly found parameters in digital reverberation units, including the following:

- Decay time
- Predelay time
- Reverberation level (mix)
- Combinations of two or more parameters at a time

The main task in the exercises and tests is to duplicate sonically a reference sound scene by listening and comparing your answer to the reference and making the appropriate changes

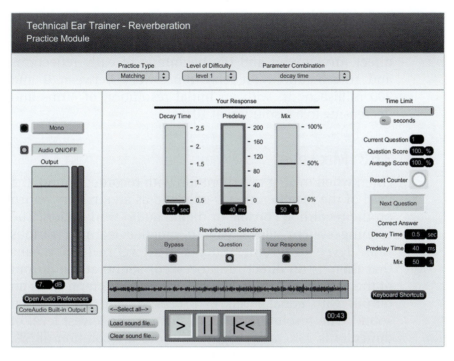

Figure 3.4 A screenshot of the user interface for the artificial reverberation trainer.

to the parameters. The software randomly chooses a parameter value based on the level of difficulty and parameter being tested, and it asks you to identify the reverberation parameters of the reference by adjusting the appropriate parameter to the value that most closely matches the sound of the reference. You can toggle between the reference question and your answer either by clicking on the switches labeled "Question" and "Your Response" (see Fig. 3.4) or by pressing the space bar on the computer keyboard. Once the two sound scenes are matched, you can click on the check answer or hit the Enter key to submit the answer and see the correct answer. Clicking on the next button moves on to the next question.

3.6.1 Decay Time

Decay times range from 0.5 seconds to 2.5 seconds with an initial resolution of 1.5 seconds and increasing in difficulty to a resolution of 0.25 seconds.

3.6.2 Predelay Time

Predelay time is the amount of time delay between the direct (dry) sound and the beginning of early reflections and reverberation. Predelay times vary between 0 and 200 milliseconds, with an initial resolution of 40 ms and decreasing to a resolution of 10 ms.

3.6.3 Mix Level

Often when mixing reverberation with recorded sound, the level of the reverberation is adjusted as an auxiliary return on the recording console or digital audio workstation. The training system allows you to practice learning various "mix" levels of reverberation. A mix level of 100% means that there is no direct (unprocessed) sound at the output of the algorithm, whereas a mix level of 50% represents an output with equal levels of processed and unprocessed sound. The resolution of mix values at the lowest level of difficulty is 25% and progresses up to a resolution of 5%, covering the range from 0 to 100% mix.

3.7 Mid-Side Matrixing

Michael Gerzon (1986, 1994) has presented mathematical explanations of matrixing and shuffling of stereo recordings to enhance and rebalance correlated and decorrelated components of a signal. His suggested techniques are useful for technical ear training because they can help in the analysis and deconstruction of a recording by bringing forth components of a sound image that might not otherwise be as audible.

By applying principles of the stereo mid-side microphone technique to completed stereo recordings, it is possible to rebalance aspects of a recording and learn more about techniques used in a recording. Although this process takes its name from a specific stereo microphone technique, any stereo recording can be postprocessed to convert the left and right channels to mid (M) and side (S), regardless of the mixing or microphone technique used.

Mastering engineers sometimes split a stereo recording into its M and S components and then process them in some way and convert them back into L and R once more.

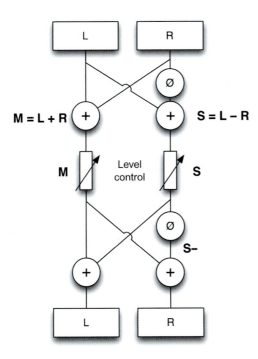

Figure 3.5 A block diagram of a mid-side matrix that allows relative adjustment of mid and side levels from left and right, and subsequent mixing back into left and right channels. We can define the mid-side matrix as mid = left + right and side = left − right.

The mid component can be derived by adding the left and right channels together. Practically, this can be done by bringing the two audio channels in on two faders and panning them both to the center. The L and R channels can be split and sent into two other pairs of channels. One pair can be panned hard left and with the L channel opposite polarity. The final pair of L and R channels can be panned right with the right channel opposite polarity. See Figure 3.5 for details on the signal routing information. Now that the signals are split into M and S, we can simply rebalance these two components, or we can apply processing to them independently. The S signal represents the components of the signal that meet either of the following conditions:

- Exist in only the L channel or only the R channel
- Are opposite of polarity, L relative to R

3.7.1 The Mid Component

The mid signal represents all components from a stereo mix that are not opposite polarity between the two channels—that is, anything that is common to both channels or just present in one side. As we can see from the block diagram presented in Figure 3.5, the M component is derived from L + R.

3.7.2 The Side Component

The side signal is derived by subtracting the L and R channels: side = L − R. Anything that is common to both L and R will be canceled out and will not form part of the S component. Any signal that is panned center in a mix will be canceled from the S component.

3.7.3 Exercise: Listening to Mid-Side Processing

The included practice module "TETlisteningMidSide" offers an easy way to audition mid and side components of any stereo recording (AIFF or WAV file formats) and hear what it sounds like if they are rebalanced. By converting a stereo mix (L and R) into M and S signals, it becomes possible to hear elements of the mix that may have been masked in the full mix. Besides being able to hear stereo reverberation better, sometimes other artifacts become apparent. Artifacts such as punch-ins, distortion, dynamic range compression, and edits can become more audible as we listen to only the S component. Many stereo mixes have a strong center component, and when that component is removed anything in the center of the stereo image is also removed. Punch-ins that are usually more of an issue with analog tape recordings are more audible when listening to the S component in isolation. A punch-in is usually performed during an overdub of a multitrack recording, where solo instrument or voice will record a part and may want to fix a certain section of the music. A punch-in is the pressing of the record button on the tape recorder for a specific track somewhere in the middle of the musical piece.

By splitting a stereo mix into its M and S components, some of the differences created by the perceptual encoding process (e.g., MP3 or AAC that has been converted back to AIFF or WAV) can be highlighted. Although the artifacts are mostly masked by the stereo audio, removing the M component does make the artifacts more audible.

Furthermore, by listening to 100% side component, we are hearing a correlation of −1 because one loudspeaker is producing the original side component and the other loudspeaker is producing an opposite polarity version of the side component.

Summary

This chapter covers the spatial attributes of sound, focusing primarily on reverberation and mid-side processing. The goal of the reverberation software practice module is to systematically familiarize listeners with aspects of artificial reverberation and to increase auditory sensitivity to time-based effects processing. By comparing two audio scenes by ear, a listener can match one or more parameters of artificial reverberation to a reference randomly chosen by the software. Listeners can progress from comparisons using percussive sound sources and coarse resolution between parameter values to more steady-state musical recordings and finer resolution between parameter values. Often very minute changes in reverberation parameters can have a significant influence on the depth, blend, spaciousness, and clarity of the final mix of a sound recording.

DYNAMIC RANGE CONTROL

CHAPTER OUTLINE

Audio Production and Critical Listening. DOI: 10.1016/B978-0-240-81295-3.00004-6

Achieving an appropriate balance of a musical ensemble is essential for expressing an artist's musical intention. Conductors and composers understand the idea of finding optimal ensemble balance for each performance and piece of music. If an instrumental part within an ensemble is not loud enough to be heard clearly, listeners do not receive the full impact of a piece of music. Overall balance depends on the control of individual vocal and instrumental amplitudes in an ensemble.

When recording spot microphone signals on multiple tracks and mixing those tracks, an engineer has some amount of control over musical balance and therefore also musical expression. When mixing multiple tracks, it can be necessary to continually adjust the level of certain instruments or voices for consistent balance from the beginning to the end of a track.

Dynamic range in the musical sense describes the difference between the loudest and quietest levels of an audio signal. For microphone signals that have a wide dynamic range, adjusting fader levels over time can compensate for variations in signal level and therefore maintain a consistent perceived loudness. Fader level adjustments made across the duration of a piece amount to manual dynamic range compression; an engineer is manually reducing the dynamic range by boosting levels during quiet sections and attenuating loud sections. Dynamic range controllers—compressors and expanders—adjust levels automatically based on an audio signal's level and can be applied to individual audio tracks or to a mix as a whole.

One type of sound that can have an extremely wide dynamic range is a lead vocal, especially when recorded with a closely placed microphone. In extreme cases in pop and rock music, a singer's dynamic range may vary from the loudest screaming to just a whisper, all within a single song. If a vocal track's fader is set to one level and left for the duration of a piece with no compression, there will be moments when the vocals will be much too loud and other moments when they will be too low. When a vocal level rises too high it becomes uncomfortable for a listener who may then want to turn the entire mix down. In the opposite situation, a vocal that is too low in level becomes difficult to understand, leaving an unsatisfying musical experience for a listener. Finding a satisfactory static fader level without compression for a sound source as dynamic as pop vocals is likely to be impossible. One way of compensating for a wide dynamic range is to manually adjust the fader level for each word or phrase that a singer sings. Although some tracks do call for such detailed manual control of fader level, use of compression is still helpful in getting part of the way to the goal of consistent, intelligible, and musically satisfying levels, especially for tracks with a wide dynamic range. Consistent levels for instruments and vocals help communicate the musical intentions of an artist more effectively.

At the same time, engineers also understand that dynamic contrast is important to help convey musical emotion. It begs the question, if the level of a vocal track is adjusted so that the fortissimo passages are the same loudness as the pianissimo passages, how is a listener going to hear any dynamic contrast? The first part of the answer to this question is that the application of level control partly depends on genre. Most classical music recordings are not going to benefit from this kind of active level control as much. For most other genres of music, at least some amount of dynamic range control is desirable. And specifically for pop and rock recordings, a more limited dynamic range is the goal so as to be consistent with recordings in this style.

Fortunately, the perception of dynamic range will remain because of timbre changes between quiet and loud dynamic levels. For almost all instruments, including voice, there is a significant increase in the number and strength of higher frequency harmonics as dynamic level goes from quiet to loud. So even if the dynamic range of a dynamic vocal performance is heavily compressed, the perception of dynamic range remains because of changes in timbre in the voice. Regardless of timbre differences, it is still possible to take dynamic range reduction too far, leaving a musical performance lifeless. Engineers still want to be aware of using too much compression and limiting because it can be fairly destructive when used excessively. Once a track is recorded with compression, there is no way to completely undo the effect. Some types of audio processing such as reciprocal peak/dip equalization allow the undoing of minor alterations with equal parameter and opposite gain settings, but compression and limiting do not offer such transparent flexibility.

Dynamic range control can be thought of as a type of amplitude modulation where the rate of modulation depends on an audio signal's amplitude envelope. Dynamics processing is simply a gain reduction applied to a signal where the gain reduction varies over time based on variations in a signal's level, with the amount of reduction based on a signal level's amplitude above a given threshold. Compression and expansion are examples of nonlinear processing because the amount of gain reduction applied to a signal depends on the signal level itself and the gain applied to a signal changes over time. Dynamics processing such as compression, limiting, expansion, and gating all offer means to sculpt and shape audio signals in unique and time-varying ways. It is time varying because the amount of gain reduction varies over time. Dynamic range control can help in the mixing process by not only smoothing out audio signal levels but by acting like a glue that helps add cohesion to various musical parts in a mix.

4.1 Signal Detection in Dynamics Processors

Dynamics processors work with objective audio signal levels, usually measured in decibels. The first reason for measuring in decibels is that the decibel is a logarithmic scale that is comparable to the way the human auditory system interprets changes in loudness. Therefore, the decibel as a measurement scale seems to correlate to the perception of sound because of its logarithmic scale. The second main reason for using decibels is to scale the range of audible sound levels to a more manageable range. For instance, human hearing ranges from the threshold of hearing, at about 0.00002 Pascals, to the threshold of pain, around 20 Pascals, a range that represents a factor of 1 million. Pascals are a unit of pressure that measure force per unit area, and are abbreviated as Pa. When this range is converted to decibels, it scales from 0 to 120 dB sound pressure level (SPL), a much more meaningful and manageable range.

To control the level of a track, there needs to be some way of measuring and indicating the amplitude of an audio signal. As it turns out, there are many ways to meter a signal, but they are all typically based on two common representations of audio signal level: peak level and RMS level (which stands for root-mean-square level). Peak level simply indicates the highest amplitude of a signal at any given time. A commonly found peak level indicator is a meter on a digital recorder, which informs an engineer how close a signal is to the digital clipping point.

The RMS is somewhat like an average signal level, but it is not mathematically equivalent to the average. With audio signals where there is a voltage that varies between positive and negative values, a mathematical average calculation is not going to give any useful information because the average will always be around zero. The RMS, on the other hand, will give a useful value and is basically calculated by squaring the signal, taking the average of some predefined window of time, and then taking the square root of that. For sine tones the RMS is easily calculated because it will always be 3 dB below the peak level or 70.7% of the peak level. For more complex audio signals such as music or speech, the RMS level must be measured directly from a signal and cannot be calculated by subtracting 3 dB from the peak value. Although RMS and average are not mathematically identical, RMS can be thought of as a type of signal average, and we will use the terms *RMS* and *average* interchangeably. Figures 4.1, 4.2, and 4.3 illustrate peak, RMS, and crest factor levels for three different signals.

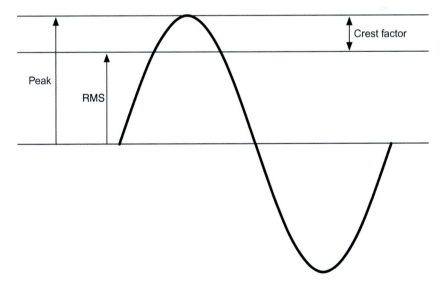

Figure 4.1 The RMS value of a sine wave is always 70.7% of the peak value, which is the same as saying that the RMS value is 3 dB below the peak level. This is only true for a sine wave. The crest factor is the difference between the peak and RMS levels, usually measured in dB. A sine wave has a crest factor of 3 dB.

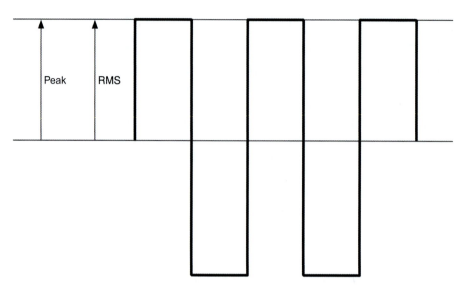

Figure 4.2 A square wave has equal peak and RMS levels, so the crest factor is 0.

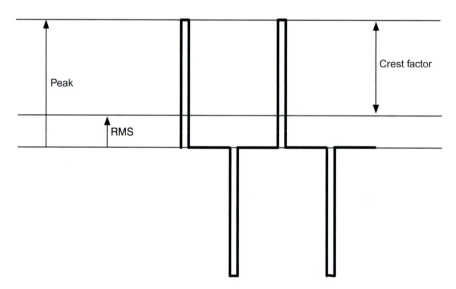

Figure 4.3 A pulse wave is similar to a square wave except that we are shortening the amount of time the signal is at its peak level. The length of the pulse determines the RMS level, where a shorter pulse will give a lower RMS level and therefore a larger crest factor.

The dynamic range can have a significant effect on the *loudness* of recorded music. The term *loudness* is used to describe the perceived level rather than the physical, measured sound pressure level. A number of factors contribute to perceived loudness, such as power spectrum and crest factor (the ratio of the peak level to the RMS level). Given two musical recordings with the same peak level, the one with a smaller crest factor will generally sound louder because its RMS level is higher. When judging the loudness of sounds, our ears respond more to average levels than to peak levels.

Dynamic range compression increases the average level through a two-stage process starting with a gain reduction of the loudest or peak levels followed by a linear output gain, sometimes called *makeup gain*. Compression and limiting essentially lower only the peaks (the loudest parts) of an audio signal and then apply a linear gain stage to bring the entire audio signal back up so that the peaks are at the maximum possible level for our recording medium (e.g., 0 dB full scale [dBFS] for digital audio). The linear gain stage after compression is sometimes called makeup gain because it is making up for peak level reduction, and some compressors and limiters apply an automatic makeup gain at the output stage. The process of compression and limiting reduces the crest factor of an audio signal, and when makeup gain is applied to restore the peaks to their original level, the RMS level is increased as well, making the overall signal louder. So by reducing the crest factor through compression and limiting, it is possible to make an audio signal sound louder even if its peak level is unchanged.

It may be tempting for a novice engineer to *normalize* a recorded audio signal in an attempt to make it sound louder. Normalizing is a process whereby a digital audio editing program scans an audio signal, finds the highest signal level for the entire clip, calculates the difference in dB between the maximum recordable level (0 dBFS) and the peak level of an audio signal, and then raises the entire audio clip by this difference so that the peak level will reach 0 dBFS. Because engineers typically want to record audio signals so that the peak levels are as close as possible to 0 dBFS, they may only get a couple of decibels of gain at best by normalizing an audio signal. This is one reason why the process of digitally normalizing a sound file will not necessarily make a recording sound significantly louder. Engineers can, however, still make a signal seem louder through the use of compression and limiting, even if the peaks are already hitting 0 dBFS.

In addition to learning how to identify the artifacts produced by dynamic range compression, it is also important to learn how to identify static changes in gain. If the overall level of a recording is increased, it is important to be able to recognize the amount of gain change applied in decibels.

4.2 Compressors and Limiters

To reduce the dynamic range of a recording, dynamics processing is used in the form of compressors and limiters. Typically a compressor or limiter will attenuate the level of a signal once it has reached or gone above a threshold level.

Compressors and expanders belong to a group of sound processing effects that are adaptive, meaning that the amount or type of processing is determined by some component of the signal itself (Verfaille et al., 2006). In the case of compressors and expanders, the amount of gain reduction applied to a signal is dependent on the level of the signal itself or a secondary signal known as a *side-chain* or *key input*. With other types of processing such as equalization and reverberation, the type, amount, or quality of processing remains the same, regardless of the input signal characteristics.

Depending on the nature of the signal-dependent processing, it can sometimes be more obvious and sometimes less obvious than processing that is not signal dependent. Any changes in processing occur synchronously with changes in the audio signal itself, and it is possible that the actual signal will mask these changes or our auditory system will assume that they are part of the original sound (as in the case of compression). Alternatively, with signal-dependent quantization error at low bit rates, the distortion (error) will be modulated by the amplitude of the signal and therefore be more noticeable than constant amplitude noise such as dither, as we will discuss in Section 5.2.3.

To determine if a signal level is above or below a specified threshold, a dynamics processor must use some method of determining the signal level, such as RMS or peak level detection.

Other forms of dynamic processing increase the dynamic range by attenuating lower amplitude sections of a recording. These types of processors are often referred to as expanders or gates. In contrast to a compressor, an expander attenuates the signal when it is *below* the threshold level. The use of expanders is common when mixing drums for pop and rock music. Each component of a drum kit is often close miked, but there is still some "leakage" of the sound of adjacent drums into each microphone. To reduce this effect, expanders or gates can be used to attenuate a microphone signal between hits on its respective drum.

There are many different types of compressors and limiters, and each make and model has its own unique "sound." This sonic signature is based on a number of factors such as the signal detection circuit or algorithm used to determine the level of an input audio signal and therefore whether to apply dynamics processing or not, and how much to apply based on the parameters set by the engineer. In analog processors, the actual electrical components in the audio signal chain and power supply also affect the audio signal.

A number of parameters are typically controllable on a compressor. These include threshold, ratio, attack time, release time, and knee.

4.2.1 Threshold

An engineer can usually set the threshold level of a compressor, although some models have a fixed threshold level with a variable input gain. A compressor starts to reduce the gain of an input signal as soon as the amplitude of the signal itself or a side-chain input signal goes above the threshold. Compressors with a side-chain or key input can accept an alternate signal input that is analyzed in terms of its level and is used to determine the gain function to be applied to the main audio signal input. Compression to the input signal is triggered when the side-chain signal rises above the threshold, regardless of the input signal level.

4.2.2 Attack Time

Although the compressor begins to reduce the gain of the audio signal as soon as its amplitude rises above the threshold, it usually takes some amount of time to achieve maximum gain reduction. The actual amount of gain reduction applied depends on the ratio and how far the signal is above the threshold. In practice, the attack time can help an engineer define or round off the attack of a percussive sound or the beginning of a musical note. With appropriate adjustment of attack time, an engineer can help a pop or rock recording sound more "punchy."

4.2.3 Release Time

The release time is the time that it takes for a compressor to stop applying gain reduction after an audio signal has gone below the threshold. As soon as the signal level falls below the threshold, the compressor begins to return it to unity gain and reaches unity gain in the amount of time specified by the release time.

4.2.4 Knee

The knee describes the transition of level control from below the threshold (no gain reduction) to above the threshold (gain reduction). A smooth transition from one to the other is called a *soft knee*, whereas an abrupt change at the threshold is known as a *hard knee*.

4.2.5 Ratio

The compression ratio determines the amount of gain reduction applied once the signal rises above the threshold. It is the ratio of input level to output level in dB above the threshold. For instance, with a 2:1 (input:output) compression ratio, the portion of the output signal that is above the threshold will be half the level (in dB) of the input signal that is above the threshold in dB. Compressors set to ratios of about 10:1 or higher are generally considered to be limiters. Higher ratios are going to give more gain reduction when a signal goes above threshold, and therefore the compression will be more apparent.

4.2.6 Level Detection Timing

To apply a gain function to an input signal, dynamics processors need to determine the amplitude of an audio signal and compare that to the threshold set by an engineer. As mentioned earlier, there are different ways to measure the amplitude of a signal, and some compressors allow an engineer to switch between two or three options. Typically the options differ in how fast the level detection is responding to a signal's level. For instance, peak level detection is good for responding to steep transients, and RMS level detection responds to less transient signals. Some dynamics processors (such as the GML 8900 Dynamic Range Controller) have fast and slow RMS detection settings, where the fast RMS averages over a shorter period of time and thus responds more to transients.

When a compressor is set to detect levels using slow RMS, it becomes impossible for the compressor to respond to very short transients. Because RMS detection is averaging over time, a steep transient will not have much influence on the averaged signal level.

4.2.7 Visualizing the Output of a Compressor

To fully understand the effect of dynamics processing on an audio signal, we need to look beyond just the input/output transfer function that is commonly seen with explanations of dynamics processors. It can be helpful to visualize how a compressor's output changes over time given a specific type of signal and thus take into account the ever-critical parameters known as *attack* and *release* time. Dynamics processors change the gain of an audio signal over time so they can be classified as nonlinear time-varying devices. They are considered nonlinear because compressing the sum of two signals is generally going to result in something different from compressing the two signals individually and subsequently adding them together (Smith, accessed August 4, 2009).

To view the effect of a compressor on an audio signal, a *step function* is required as the input signal. A step function is a type of signal that instantaneously changes its amplitude and stays at the new amplitude for some period of time. By using a step function, it is possible to illustrate how a compressor responds to an immediate change in the amplitude of an input signal and eventually settles to its target gain.

For the following visualizations, an amplitude-modulated sine wave acts as a step function (see Fig. 4.4a). The modulator is a square wave with a period of 1 second. The peak amplitude of the sine wave was chosen to switch between 1

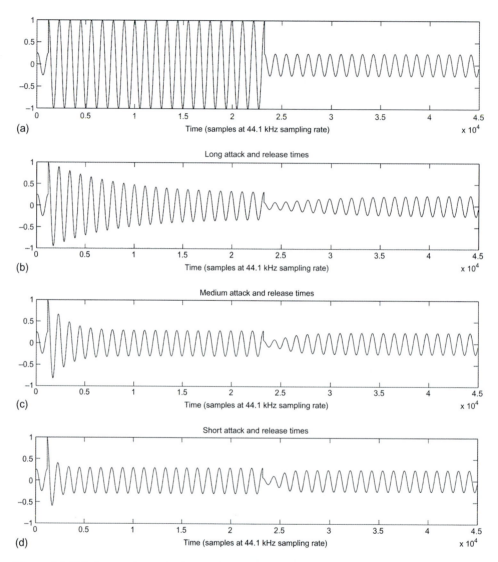

Figure 4.4 This figure shows the input signal to a compressor (a) which is an amplitude-modulated sine wave and the output of the compressor showing the step response for three different attack and release times: long (b), medium (c), and short (d).

and 0.25. An amplitude of 0.25 is 12 dB below an amplitude of 1.

Figure 4.4 shows the general attack and release curves found on most compressors. This type of visualization is not published with a compressor's specifications, but we can visualize it by recording the output when we send an

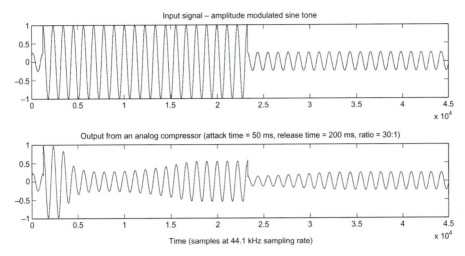

Figure 4.5 The same modulated 40-Hz sine tone through a commercially available analog compressor with an attack time of approximately 50 ms and a release time of 200 ms. Note the difference in the curve from Figure 4.4. There appears to be an overshoot in the amount of gain reduction in the attack before it settles into a constant level. A visual representation of a compressor's attack and release times such as this is not something that would be included in the specifications for a device. The difference that is apparent between Figures 4.4 and 4.5 is typically something that an engineer would listen for but could not visualize without doing the measurement.

amplitude modulated sine tone as an input signal. If this type of measurement was conducted on various types of analog and digital compressors, it would be found that they look similar in shape to what we see in Figure 4.4. Some compressor models have attack and release curves that look a bit different, such as in Figure 4.5. In this compressor there appears to be an overshoot in the amount of gain reduction in the attack before it settles into a constant level. Figure 4.6 shows an example of an audio signal that has been processed by a compressor and the resulting gain function that the compressor derived, based on the input signal level and compressor parameter settings. The gain function shows the amount of gain reduction applied over time, which varies with the amplitude of the audio signal input. The threshold was set to 6 dB, which corresponds to 0.5 in audio signal amplitude, so every time the signal goes above 0.5 in level (-6 dB), the gain function shows a reduction in level.

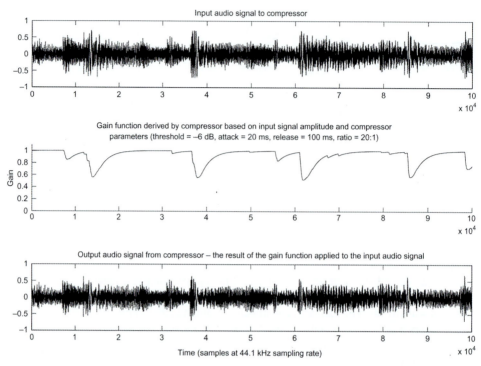

Figure 4.6 From an audio signal *(top)* sent to the input of a compressor, a gain function *(middle)* is derived based on compressor parameters and signal level. The resulting audio signal output *(bottom)* from the compressor is the input signal with the gain function applied to it. The gain function shows the amount of gain reduction applied over time, which varies with the amplitude of the audio signal input. For example, a gain of 1 (unity gain) results in no change in level, and a gain of 0.5 reduces the signal by 6 dB. The threshold was set to −6 dB, which corresponds to 0.5 in audio signal amplitude, so every time the signal goes above 0.5 in level (−6 dB), the gain function shows a reduction in level.

4.2.8 Automated Level Control through Compression

Dynamic range compression may be one of the most difficult types of processing for the beginning engineer to learn how to hear and use. Likely it is difficult to hear because often the goal of compression is to be transparent. Engineers employ a compressor when to remove amplitude inconsistencies in an instrument or voice. Depending on the nature of the signal being compressed and the parameter settings chosen, compression can range from being highly transparent to entirely obvious.

Perhaps another reason why novice engineers find it difficult to identify compression is that nearly all recorded sound that listeners hear has been compressed to some extent. Compression has become such an integral part of almost all music heard through loudspeakers that listeners can come to expect it to be part of all musical sound. Listening to acoustic music without sound reinforcement can help in the ear training process to refresh a perspective and remind oneself what music sounds like without compression.

Because dynamics processing is dependent on an audio signal's variations in amplitude, the amount of gain reduction varies with changes in the signal. With amplitude modulation of an audio signal synchronized with the amplitude envelope of the audio signal itself, the modulation can be difficult to hear because it is not clear if the modulation was part of the original signal or not. Amplitude modulation becomes almost inaudible when it reduces signal amplitude at a rate equivalent but opposite to the amplitude variations in an audio signal. Compression or limiting can be made easier to hear by setting the parameters of a device to their maximum or minimum values—a high ratio, a short attack time, a long release time, and a low threshold.

If amplitude modulation was applied that did not vary synchronously with an audio signal, the modulation would likely be much more apparent because the resulting amplitude envelope would not correlate with what is happening in the signal and would be heard it as a separate event. For instance, with a sine wave modulator, amplitude modulation is periodic and not synchronous with any type of music signal from an acoustic instrument and is therefore highly audible. This is not to say that sine tone amplitude modulation is to always be avoided. Amplitude modulation with a sine wave can sometimes produce desirable effects on an audio signal, but with that type of processing the goal is usually to highlight the effect rather than make it transparent.

Through the action of gain reduction, compressors can create audible artifacts—that is, the timbre of a sound is changed in an unwanted way—and for other circumstances these artifacts are completely intentional and contribute meaningfully to the sound of a recording. In other situations, control of dynamic range is applied without creating any artifacts, and without changing the timbre of sounds. An engineer may want to turn down the loud parts in a way that still controls the peaks but that does not disrupt the audio signal. In either case, an engineer needs to know what the artifacts sound like to decide how much or little dynamic range control to apply to a recording. On many dynamic range controllers, the user-adjustable parameters are interrelated to a certain extent and affect how an engineer uses and hears them.

4.2.9 Manual Dynamic Range Control

Because dynamic range controllers are responding to an objective measure of signal level, peak or RMS, rather than subjective signal levels, such as loudness, it is possible that the level reduction provided by a compressor does not suit an audio signal as well as desired. The automated dynamic range control of a compressor may not be as transparent as required for a given application. The amount that a compressor is acting on an audio signal is based on how much it determines an audio signal is going above a specified threshold and as a result applies gain reduction based on objective measures of signal level. Objective signal levels do not always correspond to subjective signal levels and as a result a compressor may measure a signal to be louder than it would be perceived and therefore it may apply a greater than desired attenuation.

When mixing a multitrack recording, engineers are concerned with levels, dynamics, and balance of each track, and want to be attentive to any sound sources that get masked at any point in a piece. At a more subtle level, even if a sound source is not masked, engineers strive to find the best possible musical balance, adjusting as necessary over time and across each note and phrase of music. Intent listening helps an engineer to find the best compromise on the overall levels of each sound source. It is often a compromise because it is not likely that each note of each source will be heard perfectly clearly, even with extensive dynamic range control. If each sound source is successively turned up so that it may be heard above all others, a mix will end up with all the same issues again, so it becomes a balancing act where priorities need to be set. For instance, vocals on a pop, rock, country, or jazz recording are typically the most important element. Generally an engineer wants to make sure that each word of a vocal recording is heard clearly. Vocals are often particularly dynamic in amplitude, and the addition of some dynamic range compression can help make each word and phrase of a performance more consistent in level.

With recorded sound, an engineer can influence a listener's perspective and perception of a piece of music through the use of level control on individual sound sources. A listener can be guided through a musical performance as instruments and voices are dynamically brought to the forefront and sent farther back, as the artistic vision of a performance dictates. Automation of the level of each sound source can create a changing perspective. The listener may not be consciously aware that levels are being manipulated, and, in fact, engineers often try to make the changing of levels as transparent and musical as possible. A listener should only be able to hear that each moment of a music recording is clear and musically satisfying, not that continuous level changes are being applied to a mix. Again, engineers strive to make the effect of technology transparent to an artistic vision of the music we are recording.

4.3 Timbral Effects of Compression

In addition to being a utilitarian device for managing the dynamic range of recording media, dynamics processing has become a tool for altering the color and timbre of recorded sound. When applied to a full mix, compression and limiting can help the elements of a mix coalesce. The compressed musical parts will have what is known in auditory perception as *common fate* because their amplitude changes share some similarity. When two or more elements (e.g., instruments) in a mix have synchronously changing amplitudes, the auditory system will tend to fuse these elements together perceptually. The result is that dynamics processing can help blend elements of a mix together.

In this section we will move beyond compression as a basic tool for maintaining consistent signal levels to compression as a tool to sculpt the timbre of a track.

4.3.1 Effect of Attack Time

With a compressor set to a long attack time—in the 100-millisecond range or greater—with a low threshold and high ratio we can hear the sound plunge down in level when the input signal goes above the threshold. The audible effect of the sound being brought down at this rate is what is known as a *pumping sound* and can be most audible on sounds with a strong pulse where the signal clearly rises above the threshold and then drops below it, such as those produced by drums, other percussion instruments, and sometimes an upright acoustic bass. If any lower-level sounds or background noise is present with the main sound being compressed, a modulated background sound will be heard. Sounds that are more constant in level such as distorted electric guitar will not exhibit such an audible pumping effect.

4.3.2 Effect of Release Time

Another related effect is present if a compressor is set to have a long release time, in the 100-millisecond range or greater. Listening again with a low threshold and high ratio, be attentive for the sound to come back up in level after a strong pulse. The audible effect of the sound being brought back up in level after significant gain reduction is called *breathing* because it can sound like someone taking a breath. As with the pumping effect, you may notice the effect most prominently on background sounds, hiss, or higher overtones that ring after a strong pulse.

Although compression tends to be explained as a process that it reduces the dynamic range of an audio signal, there are ways to use a compressor that can accentuate the difference between transient peak levels and any sustained resonance that may follow. In essence, what can be achieved with compression can be similar to dynamic range expansion because peaks or strong pulses can be highlighted relative to quieter sounds that immediately follow them. It may seem completely counterintuitive to try to think of compressors performing dynamic range expansion, but in the following section we will work through what happens when experimenting with various attack times.

4.3.3 Compression and Drums

A recording with a strong pulse, such as drums or percussion, with a regularly repeating transient will trigger gain reduction in a compressor and can serve as a useful type of sound to highlight the effect of a dynamics processing. By processing a stereo mix of a full drum kit through a compressor at a fairly high ratio of 6:1, attack and release times can be adjusted to hear their effect on the sound of the drums. On a typical recording of a snare drum that has not been compressed, there is a natural attack or onset, perhaps some sustain, and then a decay. The compressor can influence all of these properties depending on how the parameters are set. The attack time has the greatest influence on the onset of the drum sound, allowing an engineer to reshape this particular feature of the sound. Increasing the attack time from a very short time to a much longer time, the initial onset of each drum hit is audibly affected. A very short attack time can remove the sense of a sharp onset. By increasing the attack time, the onset sound begins to gain prominence and may in fact be accentuated slightly when compared to the uncompressed version.

Let us explore the sonic effect on a drum kit when listening through a compressor with a low threshold, high ratio, and very short attack time (e.g., down to 0 milliseconds). With such a short attack time, transients are immediately brought down in level, nearly at the rate at which the input level rises for each transient. Where the rate of gain reduction nearly matches the rate at which a transient signal rises in level, a signal's transient nature is significantly reduced. So with very short attack times, transients are lost because the gain reduction is bringing a signal's level down at nearly the same rate that the signal was originally rising up during a transient. As a result, the initial attack of a transient signal is reduced to the level of the sustained or resonant part of the amplitude envelope. Very short attack times can be useful in some instances such as on limiters that are being used to avoid clipping. For shaping drum and percussion sounds, short attack times are quite destructive and tend to take the life out of the original sounds.

Lengthening the attack time to just a few milliseconds, a clicking sound emerges at the onset of a transient. The click is produced by a few milliseconds of the original audio passing through as gain reduction occurs, and the timbre of the click is directly dependent on the length of the attack time. The abrupt gain reduction reshapes the attack of a drum hit.

By increasing the attack time further, the onset sound begins to gain prominence relative to sustain and decay portions of the sound, and it may be more accentuated than without processing. When compressing low-frequency drums such as bass drum or kick drum, an increase in the attack time will increase the presence of low-frequency harmonics. Because low frequencies have longer periods, a longer attack time will allow more cycles of a low-frequency sound to occur before gain reduction and therefore low-frequency content to be more audible on each rhythmic bass pulse.

The release time affects mostly the decay of the sound. The decay portion of the sound is that which becomes quieter after the loud onset. If the release time is set long, then the compressor gain does not quickly return to unity after the signal level has fallen below the threshold (which happens during the decay). With a long release time, the natural decay of the drum sound becomes significantly reduced. When compressing a mix of an entire drum kit, it becomes more apparent that the attack time is affecting the spectral balance of the total sound. Increasing the attack time from a very short value to something longer, increases the low-frequency energy coming from the bass drum. As attack time is lengthened from near zero to several tens or hundreds of milliseconds, the spectral effect is similar to adding a low-shelf filter to the mix and increasing the low-frequency energy.

4.3.4 Compression and Vocals

Because vocal performances tend to have a wide dynamic range, engineers often find that some sort of dynamic range control helps them reach their artistic goals for a given recording. Compression can be very useful in reducing the dynamic range and de-essing a vocal track. Unfortunately, compression does not always work as transparently as desired, and artifacts from the automated gain control of a compressor sometimes come through.

A couple of simple tips can help reduce dynamic range without adding too many of the side effects that can detract from a performance:

- *Use low ratios.* The lower the ratio, the less gain reduction that will be applied. Ratios of 2:1 are a good place to start.

- *Use more than one compressor in series.* By chaining two or three compressors in series on a vocal, each set to a low ratio, each compressor can provide some gain reduction and the effect is more transparent than using a single compressor to do all of the gain reduction.

To help identify when compression is applied too aggressively, listen for changes in timbre while watching the gain reduction meter on our compressor. If there is any change in timbre that is synchronized with gain reduction, the solution may be to lower the ratio or raise the threshold or both. Sometimes a track may sound slightly darker during extreme gain reduction, and it can be easier to identify synchronous changes when watching the gain reduction meter of a compressor.

A slight popping sound at the start of a word or phrase may indicate that the attack time is too slow. Generally a very long attack time is not effective on a vocal since it has the effect of accentuating the attack of a vocal and can be distracting.

Compression of a vocal usually brings out lower-level detail in a vocal performance such as breaths and "s" sounds. A de-esser, which can reduce the "s" sound, is simply a compressor that has a high-pass filtered (around 5 kHz) version of the vocal as its side chain or key input. De-essers tend to work most effectively with very fast attack and release times.

4.4 Expanders and Gates

4.4.1 Threshold

Expanders modify the dynamic range of an audio signal by attenuating it when its level falls *below* some predefined threshold, as opposed to compressors, which act on signal levels *above* a threshold. Gates are extreme versions of expanders and usually mute a signal when it drops below a threshold. Figure 4.7 shows the effect of an expander on an amplitude modulated sine wave. Similar to compressors, expanders often have sidechain inputs that can be used to control an audio signal with a secondary signal. For instance, engineers sometimes gate a low frequency sine tone (around 40 or 50 Hz) and with a kick drum signal sent to the sidechain input of the gate. This results in the sine

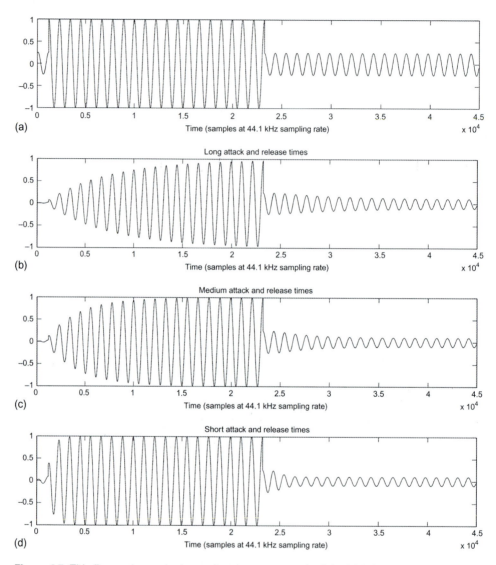

Figure 4.7 This figure shows the input signal to an expander (a) which is an amplitude modulated sine wave and the output of the expander showing the step response for three different attack and release times: short (b), medium (c), and long (c).

tone sounding only when the kick drum sounds and the two can be mixed together to create a new timbre.

Most of the controllable parameters on an expander are similar in function to a compressor with a couple of exceptions: attack and release times. These two parameters need to be considered in relation to an audio signal's level, rather than in relation to gain reduction.

4.4.2 Attack Time

The attack time on an expander is the amount of time it takes for an audio signal to return to its original level once it has gone above the threshold. Like a compressor, the attack time is the amount of time it takes to make a gain change after a signal goes above the threshold. In the case of a compressor, a signal is attenuated above threshold; with an expander, a signal returns to unity gain above threshold.

4.4.3 Release Time

The release time on an expander is the amount of time it takes for complete attenuation of an audio signal once it has dropped below the threshold. In general for compressors and expanders, release time does not define a particular direction of level control—boost or cut—it is defined with respect to a signal level relative to the threshold.

4.4.4 Visualizing the Output of an Expander

Figure 4.7 shows the effect that an expander has on the amplitude of a step function; in this case, it is an amplitude modulated sine tone. Figure 4.8 shows a clip from a music recording with the gain function derived from the audio signal and parameter settings and the resulting output audio signal. Low-level sections of an audio signal are reduced even further in the expanded audio signal.

4.5 Getting Started With Practice

The recommendations on Getting Started with Practice in Section 2.3 are applicable to all of the software exercises described in the book, and the reader is encouraged to review those recommendations on frequency and duration of practice.

The overall functionality of the software modules focused on dynamics processing, "TETpracticeDyn" and "TETpracticeExp," is very similar to that of the equalization module. With the focus on dynamics there are different parameters and qualities of sound to explore than there were with equalization.

The dynamics modules allow practice with up to three test parameters at a time: attack time, release time, and

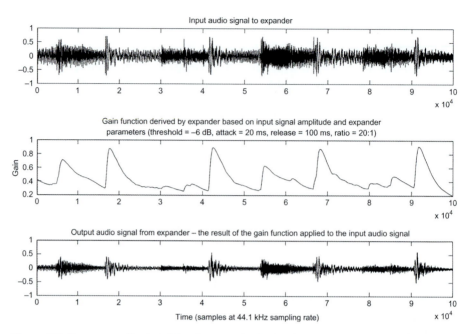

Figure 4.8 From an audio signal *(top)* sent to the input of an expander, a gain function *(middle)* is derived based on expander parameters and signal level. The resulting audio signal output *(bottom)* from the expander is the input signal with the gain function applied to it. The gain function shows the amount of gain reduction applied over time, which varies with the amplitude of the audio signal input. For example, a gain of 1 (unity gain) results in no change in level, and a gain of 0.5 reduces the signal by 6 dB. The threshold was set to −6 dB, which corresponds to 0.5 in audio signal amplitude, so every time the signal drops below 0.5 in level (−6 dB), the gain function shows a reduction in level.

ratio. Practice can occur with each parameter on its own or in combination with one or two of the other parameters, depending on what "Parameter Combination" is chosen. Threshold is completely variable for all exercises and controls the threshold for the both the computer-generated "Question" as well as "Your Response." Because the signal level of a sound recording will determine how much time a signal spends above a threshold, and it is not known how the level of every recording is going to relate to a given threshold, it is best to maintain a fully variable threshold.

In the compressor module, the threshold level should initially be set fairly low so that the effect of the compression is most audible. A make-up gain fader is included so that the subjective levels of compressed and bypassed signals can be roughly matched by ear if desired.

In the case of the expander module, a higher threshold will cause the expander to produce more pronounced changes in level. Additionally the input level can be reduced to further highlight dynamic level changes.

The Level of Difficulty option controls the number of choices available for a given parameter. With higher levels of difficulty, a greater number of parameter choices are available within each range of values.

The Parameter Combination determines which parameters will be included in a given exercise. When working with a Parameter Combination that tests only one or two parameters, the remaining user controllable parameter(s) that are not being tested will control the processing for both the "Question" and "Your Response" compressors.

The dynamic range control practice modules are the only ones of the entire collection in which the computer can choose "no compression" as a possible question. Practically this means that a ratio of 1:1 could be chosen, but only when the Parameter Combination includes "ratio" as one of the options. When a question is encountered in which no dynamic range control is heard, indicate as such by selecting a ratio of 1:1, which is equivalent to bypassing the module. If a question has a ratio of 1:1, all other parameters will be ignored in the calculation of question and average scores.

Figure 4.9 shows a screenshot of the dynamic range compression software practice module.

4.5.1 Practice Types

There are two practice types in the dynamics software practice module: Matching, Matching Memory, and Absolute Identification:

- *Matching*. Working in Matching mode, the goal is to duplicate the dynamics processing that has been applied by the software. In this mode, the user is free to switch back and forth between the "Question" and "Your Response" to determine if the dynamics processing chosen matches the unknown processing applied by the computer.

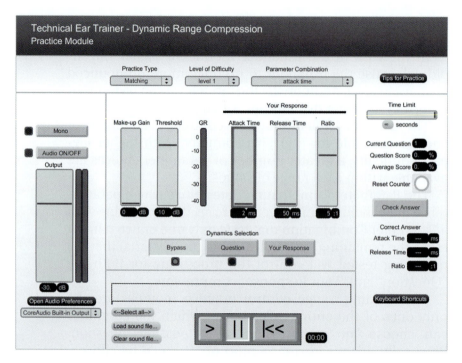

Figure 4.9 A screenshot of the software user interface for the Technical Ear Trainer practice module for dynamic range compression.

- *Matching Memory.* Similar to Matching, this mode allows free switching between "Question," "Your Response," and "Bypass" until one of the question parameters is changed. At that point, the "Question" is no longer selectable and its sound should have been memorized well enough to determine if the response is correct.
- *Absolute Identification.* This practice mode is the most difficult and requires identification of the applied dynamics processing without having the opportunity to listen to what is chosen as the correct response. You can audition only "Bypass" (no dynamics processing) and "Question" (the computer's randomly chosen processing parameters); you cannot audition "Your Response."

4.5.2 Sound Source

Any sound recording in the format of AIFF or WAV at a 44,100- or 48,000- Hz sampling rate can be used for practice. There is also an option to listen to the sound source in mono or stereo. If a sound file loaded in contains only one track of audio (as opposed to two), the audio signal will be sent out of the left output only. By pressing the mono button, the audio will be fed to both left and right output channels.

4.5.3 Recommended Recordings for Practice

A few artists are making multitrack stems available for purchase or free download. Single drum hits are useful to begin training, and then it makes sense to progress through to drum kits, as well as other solo instruments and voice. A few web sites exist with free sound samples and loops that can be used for practice, such as www.freesound. org, www.realworldremixed.com/download.php, and www. royerlabs.com, among many others. There are also excerpts or loops of various solo instruments bundled with Apple's GarageBand and Logic that can be used with the software.

Summary

This chapter discusses the functionality of compressors and expanders and their sonic effects on an audio signal. Dynamic range controllers can be used to smooth out fluctuating levels of a track, or to create interesting timbral modifications that are not possible with other types of signal processing. The compression and expansion software practice modules are described and listeners can use them to practice hearing the sonic effects of various parameter settings.

5

DISTORTION AND NOISE

Audio Production and Critical Listening. DOI: 10.1016/B978-0-240-81295-3.00005-8

In the recording process, engineers regularly encounter technical issues that cause noises to be introduced or audio signals to be degraded inadvertently. To the careful listener, such events remove the illusion of transparent audio technology, revealing a recorded musical performance and reminding them that they are listening to a recording mediated by once invisible but now clearly apparent technology. It becomes more difficult for a listener to completely enjoy any artistic statement when technological choices are adding unwanted sonic artifacts. When recording technology contributes negatively to a recording, a listener's attention becomes focused on artifacts created by the technology and drifts away from the musical performance. There are many levels and types of sonic artifacts that can detract from a sound recording, and gaining experience in critical listening promotes increased sensitivity to various types of noise and distortion.

Distortion and noise are the two broad categories of sonic artifacts that engineers typically try to avoid or use for creative effect. They can be present in a range of levels or intensities, so it is not always easy to detect lower levels of unwanted distortion or noise. In this chapter we focus on extraneous noises that sometimes find their way into a recording as well as some forms of distortion, both intentional and unintentional.

5.1 Noise

Although some composers and performers intentionally use noise for artistic effect, we will discuss some of the kinds of noise that are unwanted and therefore detract from the quality of a sound recording. Through improper grounding and shielding, loud exterior sounds, radio frequency interference, and heating, ventilation, and air conditioning (HVAC) noise, there are many sources and types of noise that engineers seek to avoid when making recordings in the studio. Frequently, noise is at a low yet still audible level and therefore will not register significantly on a meter, especially in the presence of musical audio signals.

Some of the various sources of noise include the following:
- *Clicks.* Transient sounds resulting from equipment malfunction or digital synchronization errors
- *Pops.* Sounds resulting from plosive vocal sounds
- *Ground hum and buzz.* Sounds originating from improperly grounded systems
- *Hiss, which is essentially low-level white noise.* Sounds originating from analog electronics, dither, or analog tape
- *Extraneous acoustic sounds.* Sounds that are not intended to be recorded but that exist in a recording space, such as air-handling systems or sound sources outside of a recording room

5.1.1 Clicks

Clicks are various types of short-duration, transient sounds that contain significant high-frequency energy. They may originate from analog equipment that is malfunctioning, from the act of connecting or disconnecting analog signals in a patch bay, or from synchronization errors in digital equipment interconnection.

Clicks resulting from analog equipment malfunction can often be random and sporadic, making it difficult to identify their exact source. In this case, meters can be helpful to indicate which audio channel contains a click, especially if clicks are produced in the absence of program material. A visual indication of a meter with peak hold can be invaluable to chasing down a problematic piece of equipment.

With digital connections between equipment, it is important to ensure that sampling rates are identical across all interconnected equipment and that clock sources are consistent. Without properly selected clock sources in digital audio, clicks are almost inevitable and will likely occur at some regular interval, usually spaced by several seconds. Clicks that originate from improper clock sources are often fairly subtle, and they require vigilance to identify them aurally. Depending on the digital interconnections in a studio, the clock source for each device needs to be either internal, digital input, or word clock.

5.1.2 Pops

Pops are low-frequency transient sounds that have a thump-like sound. Usually pops occur as a result of vocal plosives that are produced in front of a microphone. Plosives are consonant sounds, such as those that result from pronouncing the letters *p, b,* and *d,* in which a burst of air is produced in the creation of the sounds. A burst of air resulting from the production of a plosive arriving at a microphone capsule produces a low-frequency, thump-like sound. Usually engineers try to counter pops during vocal recording by placing a pop filter in front of a vocal microphone. Pop filters are usually made of thin fabric stretched across a circular frame.

Pops are not something heard from a singer when listening acoustically in the same space as the singer. The pop artifact is purely a result of a microphone close to a vocalist's mouth, responding to a burst of air. Pops can distract listeners from a vocal performance because they are not expecting to hear a low-frequency thump from a singer. Usually engineers can filter out a pop with a high-pass filter inserted only during the brief moment while a pop is sounding.

5.1.3 Hum and Buzz

Improperly grounded analog circuits and signal chains can cause noise in the form of hum or buzz to be introduced into analog audio signals. Both are related to the frequency of electrical alternating current (AC) power sources, which is referred to as mains frequency in some places. The frequency of a power source will be either 50 Hz or 60 Hz depending on geographic location and the power source being used. Power distribution in North America is 60 Hz, in Europe it is 50 Hz, in Japan it will be either 50 or 60 Hz depending on the specific location within the country, and in most other countries it is 50 Hz.

When a ground problem is present, there is either a hum or a buzz generated with a fundamental frequency equal to the power source alternating current frequency, 50 or 60 Hz, with additional harmonics above the fundamental. A *hum* is identified as a sound containing primarily just lower harmonics and *buzz* as that which contains more prominent higher harmonics.

Engineers want to make sure that they identify any hum or buzz before recording when the problem is easier to solve. Trying to remove such noises in postproduction is possible but will take extra time. Because a hum or buzz includes numerous harmonics of 50 or 60 Hz, a number of narrow notch filters are needed, each tuned to a harmonic, to effectively remove all of the offending sound. Although we are not going to discuss the exact technical and wiring problems that can cause hum and buzz and how such problems might be solved, there are many excellent references that cover the topic in great detail such as Giddings's book titled *Audio Systems Design and Installation* (1990).

Bringing up monitor levels while musicians are not playing often exposes any low-level ground hum that may be occurring. If dynamic range compression is applied to an audio signal and the gain reduction is compensated with makeup gain, low-level sounds including noise floor will be brought up to a more noticeable level. If an engineer can apprehend any ground hum before getting to that stage, the recording will be cleaner.

5.1.4 Extraneous Acoustic Sounds

Despite the hope for perfectly quiet recording spaces, there are often numerous sources of noise both inside and outside of a recording space that must be dealt with. Some of these are relatively constant, steady-state sounds, such as air-handling noise, whereas other sounds are unpredictable and somewhat random, such as car horns, people talking, footsteps, or noise from storms.

With most of the population concentrated in cities, sound isolation can be particularly challenging as noise levels rise and our physical proximity to others increases. Besides airborne noise there is also structure-borne noise, where vibrations are transmitted through building structures and end up producing sound in a recording space.

5.2 Distortion

Although engineers typically want to avoid or remove noises such as previously listed, distortion, on the other hand, can be used creatively as an effect, or it can appear as an unwanted artifact of an audio signal. Sometimes distortion is applied intentionally—such as to an electric guitar signal—to enhance the timbre of a sound, adding to the palette of available options for musical expression. At other times, an audio signal may be distorted through improper parameter settings, malfunctioning equipment, or low-quality equipment. Whether or not distortion is intentional, an engineer should be able to identify when it is present and either shape it for artistic effect or remove it, according to what is appropriate for a given recording.

Fortunately engineers do have an aid to help identify when a signal gets clipped in an objectionable way. Digital meters, peak meters, clip lights, or other indicators of signal strength are present on most input stages of analog-to-digital converters, microphone preamplifiers, as well as many other gain stages. When a gain stage is overloaded or a signal clipped, a bright red light provides a visual indication as soon as a signal goes above a clip level, and it remains lit until the signal has dropped below the clip level. A visual indication in the form of a peak light, which is synchronous with the onset and duration of a distorted sound, reinforces an engineer's awareness of signal degradation and to help identify if and when a signal has clipped. Unfortunately, when working with large numbers of microphone signals, it can be difficult to catch every flash of a clip light, especially in the analog domain. Digital meters, on the other hand, allow *peak hold* so that if a clip indicator light is not seen at the moment of clipping, it will continue to indicate that a clip did occur until it is reset manually by an engineer. For momentary clip indicators, it becomes that much more important to rely on what is heard to identify overloaded sounds because it can be easy to miss the flash of a red light.

In the process of recording any musical performance, engineers set microphone preamplifiers to give as high a recording level as possible, as close to the clip point as possible, but without going over. The goal is to maximize signal-to-noise or signal-to-quantization error by recording a signal whose peaks reach the maximum recordable level, which in digital audio is 0 dB full scale. The problem is that the exact peak level of a musical performance is not known until after it has happened. Engineers set preamplifier gain based on a representative sound check, giving themselves some headroom in case the peaks are higher than what is expected. When the actual musical performance occurs following a sound check, often the peak level will be higher than it was during sound check because the musicians may be performing at a more enthusiastic and higher dynamic level than they were during the sound check.

Although it is ideal to have a sound check, there are many instances in which engineers do not have the opportunity to do so, and must jump directly into recording, hoping that their levels are set correctly. They have to be especially concerned about monitoring signal levels and detecting any signal clipping in these types of situations.

There is a range of sounds or qualities of sound that we can describe as distortion in a sound recording. Among these unwanted sounds are the broad categories of distortion and noise. We can expand on these categories and outline various types of each:

- *Hard clipping or overload.* This is harsh sounding and results from a signal's peaks being squared off when the level goes above a device's maximum input or output level.
- *Soft clipping or overdrive.* Less harsh sounding and often more desirable for creative expression than hard clipping, it usually results from driving a specific type of circuit designed to introduce soft clipping such as a guitar amplifier.
- *Quantization error distortion.* Resulting from low bit quantization in PCM digital audio (e.g., converting from 16 bits per sample to 8 bits per sample). Note that we are not talking about low bit-rate perceptual encoding but simply reducing the number of bits per sample for quantization of signal amplitude.
- *Perceptual encoder distortion.* There are many different artifacts, some more audible than others, that can occur when encoding a PCM audio signal to a data-reduced version (e.g., MP3 or AAC). Lower bit rates exhibit more distortion.

There are many forms and levels of distortion that can be present in reproduced sound. All sound reproduced by loudspeakers is distorted to some extent, however insignificant. Equipment with exceptionally low distortion can be particularly expensive to produce, and therefore the majority of average consumer audio systems exhibits slightly higher levels of distortion than those used by professional audio engineers. Audio engineers and audiophile enthusiasts go to great lengths (and costs) to reduce the amount of distortion in their signal chain and loudspeakers.

Most other commonly available sound reproduction devices such as intercoms, telephones, and inexpensive headphones connected to digital music players have audible distortion. For most situations such as voice communication, as long as the distortion is low enough to maintain intelligibility, distortion is not really an issue. For inexpensive audio reproduction systems, the level of distortion is usually not detectable by an untrained ear. This is part of the reason for the massive success of the MP3 and other perceptually encoded audio formats found on Internet audio—most casual listeners do not perceive the distortion and loss of quality, yet the size of files is much more manageable and audio files are much more easily transferable over a computer network connection than their PCM equivalents.

Distortion is usually caused by amplifying an audio signal beyond an amplifier's maximum output level. Distortion can also be produced by increasing a signal's level beyond the maximum input level of an analog-to-digital converter (ADC). When an ADC attempts to represent a signal whose level is above 0 dB full scale (dB FS), called an *over*, the result is a harsh-sounding distortion of the signal.

5.2.1 Hard Clipping and Overload

Hard clipping occurs when too much gain is applied to a signal and it attempts to go beyond the limits of a device's maximum input or output level. Peak levels greater than the maximum allowable signal level of a device are flattened, creating new harmonics that were not present in the original waveform. For example, if a sine wave is clipped such as in Figure 5.1, the result is a square wave such as in Figure 5.2, whose time domain waveform now contains sharp edges and whose frequency content contains additional harmonics. A square wave is a specific type of waveform that is composed of odd numbered harmonics (1st, 3rd,

Chapter 5 DISTORTION AND NOISE **113**

5th, 7th, and so on). One of the results of distortion is an increase in the numbers and levels of harmonics present in an audio signal. Technical specifications for a device often indicate the total harmonic distortion for a given signal level, expressed as a percentage of the overall signal level.

Because of the additional harmonics that are added to a signal when it is distorted, the sound takes on an increased brightness and harshness. Clipping a signal flattens out the peaks of a waveform, adding sharp corners to a clipped peak. The new sharp corners in the time domain waveform represent increased high-frequency harmonic content in the signal, which would be confirmed through frequency domain analysis and representation of the signal.

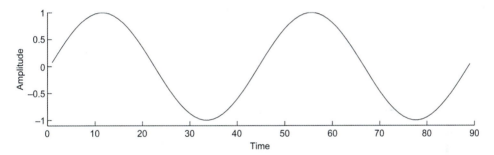

Figure 5.1 A sine wave at 1 kHz.

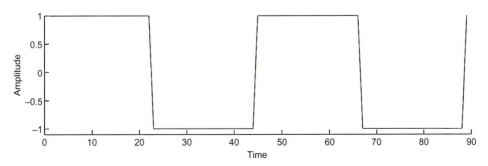

Figure 5.2 A sine wave at 1 kHz that has been hard clipped. Note the sharp edges of the waveform that did not exist in the original sine wave.

5.2.2 Soft Clipping

A milder form of distortion known as *soft clipping* or *overdrive* is often used for creative effect on an audio signal. Its timbre is less harsh than clipping, and as can be seen from Figure 5.3, the shape of an overdriven sine wave does not have the sharp corners that are present in a hard-clipped sine wave (Fig. 5.2). As is known from frequency analysis, the sharp corners and steep vertical portions of a clipped sine waveform indicate the presence of high-frequency harmonics.

Hard clipping distortion is produced when a signal's amplitude rises above the maximum output level of an amplifier. With gain stages such as solid-state microphone preamplifiers, there is an abrupt change from linear gain before clipping to nonlinear distortion. Once a signal reaches the maximum level of a gain stage, it cannot go any higher regardless of an increasing input level; thus there are flattened peaks such as in Figure 5.2. It is the abruptness of the change from clean amplification to hard clipping that introduces such harsh-sounding distortion.

In the case of soft clipping, there is a gradual transition, instead of an abrupt change, between linear gain and maximum output level. When a signal level is high enough to reach into the transition range, there is some flattening of a signal's peaks (as in Fig. 5.3) but the result is less harsh than with hard clipping.

In recordings of pop and rock music especially, there are examples of the creative use of soft clipping and overdrive that enhance sounds and create new and interesting timbres.

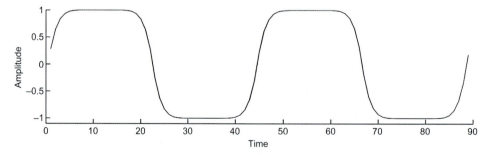

Figure 5.3 A sine wave at 1 kHz that has been soft clipped or overdriven. Note how the shape of the waveform is somewhere in between that of the original sine wave and a square wave.

5.2.3 Quantization Error Distortion

In the process of converting an analog signal to a digital PCM representation, analog amplitude levels for each sample get quantized to a finite number of steps. The number of bits of data stored per sample determines the number of possible quantization steps available to represent analog voltage levels. An analog-to-digital converter records and stores sample values using binary digits, or bits, and the more bits available, the more quantization steps possible.

The Red Book standard for CD-quality audio specifies 16 bits per sample, which represents 2^{16} or 65,536 possible steps from the highest positive voltage level to the lowest negative value. Usually higher bit depths are chosen for the initial stage of a recording. Given the choice, most recording engineers will record using at least 24 bits per sample, which corresponds to 2^{24} or 16,777,216 possible amplitude steps between the highest and lowest analog voltages. Even if the final product is only 16 bits, it is still better to record initially at 24 bits because any gain change or signal processing applied will require requantization. The more quantization steps that are available to start with, the more accurate the representation of an analog signal.

Each quantized step of linear PCM digital audio is an approximation of the original analog signal. Because it is an approximation, there will be some amount of error in any digital representation. Quantization error is essentially the distortion of an audio signal. Engineers usually minimize quantization error distortion by applying dither or noise shaping, which randomizes the error. With the random error produced by dither, distortion is replaced by constant noise which is generally considered to be preferable over distortion.

The interesting thing about the amplitude quantization process is that the signal-to-error ratio drops as signal level is reduced. In other words, the error becomes more significant for lower-level signals. For each 6 dB that a signal is below the maximum recording level of digital audio (0 dB FS), 1 bit of binary representation is lost. For each bit lost, the number of quantization steps is halved. A signal recorded at 16 bits per sample at an amplitude of -12 dB FS will only be using 14 of the 16 bits available, representing a total of 16,384 quantization steps.

Although the signal peaks of a recording may be near the 0 dB FS level, there are often other lower-level sounds within a mix that can suffer more from quantization error. Many recordings that have a wide dynamic range may include significant portions where audio signals hover at some level well below 0 dB FS. One example of low-level sound within a recording is reverberation and the sense of space that it creates. With excessive quantization error, perhaps as the result of bit depth reduction, some of the sense of depth and width that is conveyed by reverberation is lost. By randomizing quantization error with the use of dither during bit depth reduction, some of the lost sense of space and reverberation can be reclaimed, but with the cost of added noise.

5.2.4 Software Module Exercises

The included software module "TETpracticeDist," focusing on distortion, allows the listener to practice hearing three different types of distortion: soft clipping, hard clipping, and distortion from bit depth reduction.

There are two main practice types with this software module: Matching and Absolute Identification. The overall functioning of the software is similar to other modules discussed previously.

5.2.5 Perceptual Encoder Distortion

Perceptual encoding of audio significantly reduces the amount of data required to represent an audio signal with only minimal degradation in audio quality. In this section we are concerned with lossy audio data compression, which removes audio during the encoding process. There are also lossless encoding formats that reduce the size of an audio file without removing any audio. Lossless encoding is comparable to the ZIP computer file format, where file size is reduced but no actual data are removed.

By converting a linear PCM digital audio file to a data-compressed, lossy format such as MP3, 90% of the data used to represent a digital audio signal are removed and yet the encoded version still sounds similar to the original uncompressed audio file. The differences in sound quality between an encoded version of a recording and the original PCM version are mostly imperceptible by the average listener, yet these same differences in sound quality can be a huge source of frustration to an experienced sound engineer. Because of signal degradation during the encoding process, perceptual encoding is considered a type of distortion, but it is a type of distortion that is not easily measurable, at least objectively. Due to the difficulty in obtaining meaningful objective measurements of distortion and sound quality with perceptual encoders, their development has involved expert listeners who are adept at identifying audible artifacts resulting from the encoding process. Expert listeners audition music recordings encoded at various bit rates and levels of quality and then rate audio quality on a subjective scale. Trained expert listeners become adept at quickly identifying distortion and artifacts produced by perceptual encoders because they know where to focus their aural attention and for what to listen.

With the proliferation of downloadable music from the Internet, perceptually encoded music has become ubiquitous, with the most well-known version being the MP3, more technically known as MPEG-1 Audio Layer-3. There are many other encoding-decoding (codec) schemes that go by names such as AAC (Advanced Audio Coding), WMA (Windows Media Audio), AC-3 (also known as Dolby Digital), and DTS (Digital Theater Systems). Codecs reduce the amount of data required to represent a digital audio signal by removing components of a signal that are deemed to be inaudible based on psychoacoustic models. The primary improvement in codecs over years of development and progression has been that they are more intelligent in how they remove audio data and they are increasingly transparent at lower bit rates. That is, they produce fewer audible artifacts for a given bit rate than the previous generation of codecs. The psychoacoustic models that are used in codecs have become more complex, and the algorithms used in signal detection and data reduction based on these models have become more precise. Still, when compared side by side with an original, unaltered signal, it is possible to hear the difference between the two.

The process of converting a linear PCM digital audio (such as AIFF, WAV, or BWF) to an MP3, AAC, WMA, RealAudio, or other lossy encoded format removes components of an audio signal that an encoder deems we cannot hear. Encoders perform various types of analyses to determine the frequency content and dynamic amplitude envelope of an audio signal, and based on psychoacoustic models of human hearing the encoders remove components of an audio signal that are likely inaudible. Some of these components are quieter sounds that are partially masked by louder sounds in a recording. Whatever sounds are determined to be masked or inaudible are removed and the resulting encoded audio signal can be represented with fewer data than were used to represent the original signal. Unfortunately, the encoding process also removes audible components of an audio signal, and therefore encoded audio sounds are degraded relative to an original nonencoded signal.

As we explore the audible artifacts and signal distortion of encoded audio, here are some items upon which to focus while practicing critical listening:

- *Clarity and sharpness.* Listen for some loss of clarity and sharpness in percussion and transient signals. The loss of clarity can translate into a feeling that there is a thin veil covering the music. When compared to linear PCM, the unencoded audio should sound more direct.
- *Reverberation.* Listen for some loss of reverberation and other low-amplitude components. The effect of lost reverberation usually translates into less depth and width in a recording and the perceived space around the music (acoustic or artificial) is less apparent.
- *Encoded audio.* A bit gurgly or swooshy. Musical notes that are held, especially with prominent solo instruments or vocals, do not sound as smooth as they should, and the overall sound can take on a tinny quality.
- *Nonharmonic high-frequency sounds.* These sounds, such as those from cymbals, and noiselike sounds, such as an audience clap-ping, can take on a swooshy quality.

5.2.6 Exercise: Comparing Linear PCM to Encoded Audio

It is important to investigate how various perceptual encoders affect sound quality. One of the ways to explore sound quality degradation is to encode linear PCM sound files and compare the original to the encoded version to identify any audible differences. There are many free pieces of software that will encode audio signals, such as Apple's iTunes Player and Microsoft's Windows Media Player. Sound quality deficiencies in encoded audio may not be immediately obvious unless we are tuned into the types of artifacts that are produced when audio is encoded. By switching back and forth between a linear PCM audio file and an encoded version of the same audio, it becomes easier to hear any differences that may be present. Once we begin to learn to hear the kinds of artifacts an encoder is producing, they become easier to hear without doing a side-by-side comparison of encoded to linear PCM.

Start by encoding a linear PCM audio file at various bit rates in MP3, AAC, or WMA and try to identify how an audio signal is degraded. Lower bit rates result in a smaller file size, but they also reduce the quality of the audio. Different codecs—MP3, AAC, and WMA—provide slightly different results for a given bit rate because the method of encoding varies from codec to codec. Switch back and forth between the original linear PCM audio and the encoded version. Try encoding recordings from different genres of music. Note the sonic artifacts that are produced for each bit rate and encoder.

Another option is to compare streaming audio from online sources to linear PCM versions that you may have. Most online radio stations and music players are using lower-bit-rate audio containing more clearly audible encoding artifacts than is found with audio from other sources such as through the iTunes Store.

5.2.7 Exercise: Subtraction

Another interesting exercise to conduct is to subtract an encoded audio file from an original linear PCM version of the same audio file. To complete this exercise, convert a linear PCM file to some encoded form and then convert it back to linear PCM at the same sampling rate. Import the original sound file and the encoded/decoded file (now linear PCM) into a digital audio workstation (DAW), on two different stereo tracks, taking care to line them up in time as precisely as possible. By playing back the synchronized stereo tracks together, reverse the polarity of the encoded/decoded file so that it is subtracted from the original. Provided the two stereo tracks are lined up accurately in time, anything that is common to both tracks will cancel, and the remaining audio that is heard hear is that which was removed by the codec. By doing this exercise, helps highlight the types of artifacts that are present in encoded audio.

5.2.8 Exercise: Listening to Encoded Audio through Mid-Side Processing

By splitting an encoded file into its mid and side (M-S) components, some of the artifacts created by the encoding process can be uncovered. The perceptual encoding process relies on masking to hide artifacts that are created in the process. When a stereo recording is converted to M and S components and the M component is removed, artifacts typically become much more audible. In many recordings, especially in the pop/rock genre, the M component forms the majority of the audio signal and can mask a great deal of encoding artifacts. By reducing the M component, the S component becomes more audible along with encoder artifacts.

Try encoding an audio file with perceptual encoder at a common bit rate such as 128 kbps and decoding it back into linear PCM (WAV or AIFF). It is possible to use the M-S matrix software module included with this book to hear the effect that M-S decoding can have on highlighting the effects of a codec.

Summary

In this chapter we explored some of the undesirable sounds that can make their way into a recording. By practicing with the included distortion software ear-training module and completing the exercises, we can become more aware of some common forms of distortion.

AUDIO CLIP EDIT POINTS

Audio Production and Critical Listening. DOI: 10.1016/B978-0-240-81295-3.00006-X

In Chapter 4 we discussed the modification of an audio signal's amplitude envelope through dynamics processing. In this chapter we will explore the amplitude envelope and technical ear training from a slightly different perspective: from that of an audio editor.

The process of digital audio editing, especially with classical or acoustic music using a source-destination method, offers an excellent opportunity for ear training. Likewise, the process of music editing requires an engineer to have a keen ear for transparent splicing of audio. Music editing involves making transparent connections or splices between takes of a piece of music, and it often requires specifying precise edit locations by ear. In this chapter we will explore how aspects of digital editing can be used systematically as an ear training method, even out of the context of an editing session. The chapter describes a software tool based on audio editing techniques that is an effective ear trainer offering benefits that transfer beyond audio editing.

6.1 Digital Audio Editing: The Source-Destination Technique

Before describing the software and method for ear training, it is important to understand some digital audio editing techniques used with classical music. Classical music requires a high level of precision—perhaps more so than other types of music—to achieve the level of transparency required.

Empirically, through hundreds of hours of classical music editing, I have found that the process of repeatedly adjusting the location of edit points and creating smooth cross-fades by ear not only results in a clean recording but can also result in improved listening skills that translate to other areas of critical listening. Through highly focused listening required for audio editing, with the goal of matching edit points from different takes, the editing engineer is participating in an effective form of ear training.

Digital audio editing systems allow an editing engineer to see a visual representation of a waveform and to move, insert, copy, or paste audio files to any location along a visual timeline. For significant portions of music recording editing, a rough estimate of an edit location is found first, followed by the precise placement of an edit point location through listening. Despite having a visual representation of a waveform, it is often more efficient and more accurate to find the precise location of an edit by ear.

During the editing process, an engineer is given a list of takes from a recording session and assembles a complete piece of music using the best takes from each section of a musical score. A common method of editing classical or acoustic music is known as source-destination. The engineer essentially builds a complete musical performance (the *destination*) by taking the best excerpts from a list of recording session takes (the *source*) and piecing them together.

In source-destination editing, the location of an edit is found by following a musical score and placing a marker at a chosen edit point along the timeline of the visual waveform representing the recorded music. The editing engineer usually auditions a short excerpt—typically 0.5 to 5 seconds in length—of a recorded take, up to a specific musical note at which an edit is to occur. Next, the same musical excerpt from a different take is auditioned and compared to the first one. Usually the end point of such an excerpt will be chosen to occur precisely at the onset of a musical note, and therefore the point of connection will be inaudible. The goal of an editing engineer is to focus on the sonic characteristics of the note onset that occurs during the final few milliseconds of an excerpt and match the sound quality between takes by adjusting the location of the edit point (i.e., the end point of the excerpt). The edit point marker may appear as a movable bracket on the audio signal waveform, as in Figure 6.1.

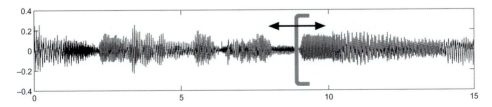

Figure 6.1 A typical view of a waveform in a digital editor with the edit point marker that indicates where the edit point will occur and the audio will cross-fade into a new take. The location of the marker, indicated by a large bracket, is adjustable in time (left/sooner or right/later). The arrow indicates simply that the bracket can slide to the left or right. The editing engineer will listen to the audio up to this large bracket with a predetermined preroll time that usually ranges from 0.5 to 5 seconds.

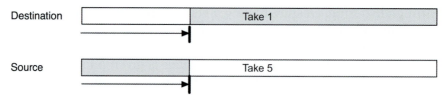

Figure 6.2 The editing engineer auditions both the source and the destination audio files, up to a chosen edit point, usually at the onset of a note or beat. In an editing session, the two audio clips (source and destination) would be of identical musical material but from different takes. The engineer auditions the audio excerpts up to a chosen edit point, usually placed midway through the attack of a note or strong beat. One of the goals of the engineer is to answer the question, does the end point in the source match that of the destination? The greater the similarity between the two cut-off timbres, the more successful the edit will be. The software module presented here re-creates the process of auditioning a sound clip up to a predefined point and matching that end point in a second sound clip.

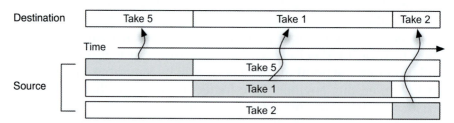

Figure 6.3 Source and destination waveform timelines are shown here in block form along with an example of how a set of takes (source) might fit together to form a complete performance (destination). In this example it is assumed that takes 1, 2, and 5 would be of the same musical program material and therefore a composite version could be produced of the best sections from each take to form what is labeled as the destination in this figure.

It is the editing engineer's focus on the final milliseconds of an audio excerpt that is critical to finding an appropriate edit point. When an edit point is chosen to be at the onset of a musical note, it is important to set the edit point such that it actually occurs sometime during the beginning of a note attack. Figure 6.1 shows a gate (square bracket indicating the edit point) aligned with the attack of a note.

When an engineer auditions a clip of audio up to a chosen edit point, the new note that begins to sound, but is immediately stopped, can form a transient, percussive sound. The specific characteristics of the actual sound of the cut note will vary directly with the amount of the incoming note that is sounded before being cut. Figure 6.2 illustrates in block form, the process of auditioning source and destination program material.

Once the characteristics of the final few milliseconds of audio are matched as closely as possible between takes, an edit is made with a cross-fade from one take into another and auditioned to check for any sonic anomalies. Figure 6.3 illustrates a composite version as the destination which has been drawn from three different source takes.

During the process of auditioning a cross-fade, an editing engineer also pays close attention to the sound quality of the cross-fade, which typically may range from a few to several hundred milliseconds depending on the context (e.g., sustained notes versus transients). The process of listening back to a cross-fade and adjusting the cross-fade parameters such as length, position, and shape also offers an opportunity to improve critical listening skills.

6.2 Software Exercise Module

Based on source-destination editing, the included ear training software module was designed to mimic the process of comparing the final few milliseconds of two short clips of identical music from different takes. The advantage of the software practice module is that it promotes critical listening skills without requiring an actual editing project. The main difference in working with the practice module is that the software will work with only one "take," that being any linear PCM sound file loaded in. Because of this difference, the two clips of audio will be identical signals, and therefore it is possible to find identical sounding end points. The benefit of working this way is that the software has the ability to judge if the sound clips end at precisely the same point.

To start, the software randomly chooses a short excerpt or clip (which is called clip 1 or the reference) from any stereo music recording loaded into the software. The exact duration of clip 1 is not revealed, but it can be auditioned. The excerpt lengths, which range from 500 milliseconds to 2 seconds, are also chosen at random to ensure that one is not simply being trained to identify the duration of the audio clips. A second clip (clip 2 or your answer) of known duration, and with a starting point identical to clip 1, can also be auditioned and compared to clip 1. Clips can be auditioned as many times as necessary by pressing the appropriate button or keyboard shortcut.

The goal of the exercise is to adjust the duration of clip 2 until it ends at exactly the same point in time as clip 1. By listening to the amplitude envelope, timbre, and musical content of the final few milliseconds of each clip, it is possible to compare the two clips and adjust the length of clip 2 so that the sound of its end point matches clip 1. By pursuing a cycle of auditioning, comparing, and adjusting the length of clip 2, the goal is to identify the end point characteristics of clip 1 and match those characteristics with clip 2.

The length of clip 2 is adjusted by "nudging" the end point either earlier or later in time. There are different nudge time step sizes from which to choose so the duration of the clip can be adjusted in increments of 5, 10, 15, 25, 50, or 100 milliseconds. The smaller the nudge step size, the more difficult it is to hear a difference from one step to another.

Figure 6.4 shows the waveforms of four sound clips of increasing length from 825 ms to 900 ms in steps of 25 ms. This particular example shows how the end of the clip can vary significantly depending on the length chosen. Although the second (850 ms) and third (875 ms) waveforms in Figure 6.4 look very similar, there is a noticeable difference in the perceived percussive or transient sound at the end. With smaller step or nudge sizes, the difference between steps would be less obvious and would require more training for correct identification.

After deciding on a clip length, the "Check Answer" button can be pressed to find out the correct answer and continue to audition the two clips for that question. The software indicates whether the response for the previous question was correct or not and if incorrect, it indicates whether clip 2 was too short or too long and the magnitude of the error. Figure 6.5 shows a screenshot of the software module.

There is no view of the waveform as would typically be seen in a digital editor because the goal is to create an environment where we must rely solely on what is heard with

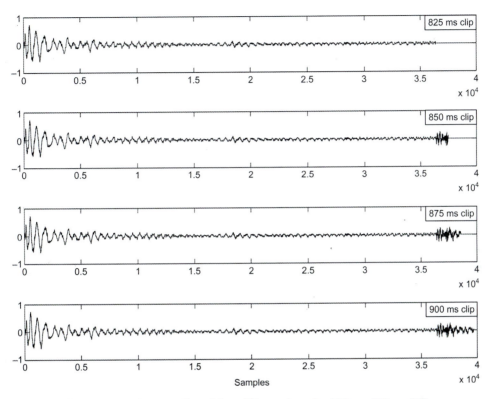

Figure 6.4 Clips of a music recording of four different lengths: 825 ms, 850 ms, 875 ms, and 900 ms. This particular example shows how the end of the clip can vary significantly depending on the length chosen. The listener can focus on the quality of the percussive sound at the end of the clip to determine the one that sounds most like the reference. The clip of 825 ms duration contains a faint percussive sound at the end of the clip, but because the note that begins to sound (a drum hit in this case) is almost completely cut off, it comes out as a short click. In this specific example, the listener can focus on the percussive quality, timbre, and envelope of the incoming drum hit at the end of the clip to determine the correct sound clip length.

minimal visual information about the audio signal. There is, however, a black bar that increases in length over a time-line, tracking the playback of clip 2 in real time, as a visual indication that clip 2 is being played. Also, the play buttons for the respective clips turn green briefly while the audio is playing and then return to gray when the audio stops.

With this ear training method, the goal is to compare one sound to another and attempt to match them. There is no

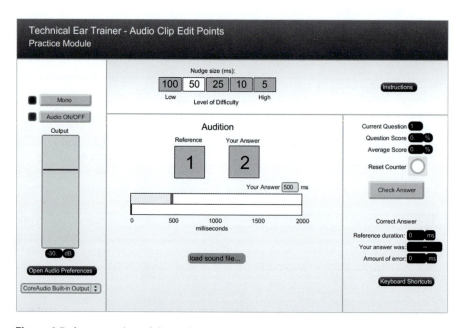

Figure 6.5 A screenshot of the training software. The large squares with "1" and "2" are playback buttons for clips 1 and 2, respectively. Clip 1 (the reference) is of unknown duration, and the length of clip 2 must be adjusted to match clip 1. Below the clip 2 play button are two horizontal bars. The top one indicates, with a vertical bar, the duration of clip 2, in the timeline from 0 to 2000 milliseconds. The bottom bar increases in length (from left to right) up to the vertical line in the top bar, tracking the playback of clip 2, to serve as a visual indication that clip 2 is being played.

need to translate the sound feature to a verbal descriptor but instead the focus lies solely on the features of the audio signal. Although there is a numeric display indicating the length of the sound clip, this number serves only as a reference for keeping track of where the end point is set. The number has no bearing on the sound features heard other than for a specific excerpt. For instance, a 600-ms randomly chosen clip will have different end point features from most other randomly chosen 600-ms clips.

Practice exercises should progress from the less challenging exercises using large step sizes of 100 ms through to the most challenging exercises where the smallest step size is 5 ms.

Almost any stereo recording in the format of linear PCM AIFF or WAV can be used with the training software, as long as it is at least 30 seconds in duration.

6.3 Focus of the Exercise

With the type of training program described in this chapter, the main goal is to focus on the amplitude envelope of a signal at a specific point in time—that being the end of a short audio excerpt. Although the audio is not being processed in any way, the location of the end point determines how and at what point a musical note may get cut. In this exercise focus on the final few milliseconds of the first clip, hold the end sound in memory, and compare it to the second clip.

Because the software randomly chooses the location of an excerpt, an end point can occur almost anywhere in an audio signal. Nonetheless, there are two specific cases where the location of a cut is important to describe: those that occur at the entrance of a strong note or beat and those that occur during a sustained note, between strong beats.

First, the outcome of a cut falling at the beginning of a strong note or beat can be explored. If the cut occurs during the attack portion of a musical note, a transient signal may be produced whose characteristics vary where a note's amplitude envelope is cut, allowing the matching of a transient sound by adjusting the cutoff point. Depending on how much of a note or percussive sound gets cut off, the spectral content of that particular sound will vary with the note's modified duration. With regard to a clipped note at the end, generally a shorter note segment will have a higher spectral centroid than a longer segment and have a brighter sound quality. The spectral centroid of an audio signal is the average frequency of a spectrum and it describes where the center of mass of a spectrum is located. If there is a click at the end of an excerpt—produced as a result of the location of the end point relative to the waveform—it can serve as a cue for the location of the end point. The spectral quality of the click can be assessed and matched based on its duration.

Next the case is examined of a more sustained or decaying audio signal that is cut. For this type of cut, focus should be placed on the duration of the sustained signal and match its length. This might be analogous to adjusting the hold time of a gate (dynamics processor) with a very short release time. With this type of matching, the focus may shift more to musical qualities such as tempo in determining how long the final note is held before being muted.

With any end point location, the requirement is to track the amplitude envelope and spectral content of the end of the clip. One goal of this exercise is to increase hearing acuity, facilitating the ability to hear subtle details in a sound recording that were not apparent before spending extensive time doing digital editing. Practicing with this exercise may begin to highlight details of a recording that may not have been as apparent when the entire musical piece was auditioned. By listening to short excerpts out of context of the musical piece, sounds within a recording can be heard in new ways, and some sounds may become unmasked and thus more audible. It allows a focus on features that may be partially or completely masked when heard in context (i.e., much longer excerpts) or features that are simply less apparent in the larger context. Repetition of the clips out of context of the entire recording may also contribute to a change in the perception of an audio signal. It is common for music composers to take excerpts of musical recordings and repeat them to create a new type of sound and effect, allowing listeners to hear new details in the sound that may not have been apparent before.

The ear training method can help us focus on quieter or lower-level features (in the midst of louder features) of a given program material. Quieter features of a program are those features that may be partly or mostly masked, perceptually less prominent, or considered in the background of a perceived sound scene or sound stage. Examples might include the following (those listed earlier are included here again):

- Reverberation and delay effects for specific instruments
- Artifacts of dynamic range compression for specific instruments
- Specific musical instrument sound quality: a drummer's brush sounds or the articulation of acoustic double bass on a jazz piece
- Specific features of each musical voice/instrument such as the temporal nature or spatial location of amplitude envelope components (attack, decay, sustain, and release)
- Definition and clarity of elements within the sound image, width of individual elements

Sounds taken out of context start to give a new impression of the sonic quality and also the musical feel of a recording. Additional detail from an excerpt is often heard when a short clip of music is played repeatedly, detail that would not necessarily be heard in context.

Working with this practice module and a musical example that features prominent vocals, acoustic bass, acoustic guitar, piano, and drums played lightly (such as "Desafinado" by Stan Getz and João Gilberto [1963]), brings forth new impressions of the timbres and sound qualities found in the recording that were not apparent previously.

In this recording, the percussion part is fairly quiet and more in the background, but if an excerpt falls between vocal phrases or guitar chords, the percussion part may perceptually move to the foreground as the matching exercise changes our focus. It also may be easier to focus on characteristics of the percussion, such as its reverberation or echo, if that particular musical part can be heard more clearly. Once details are identified within a small excerpt, it can make it easier to hear these features within the context of the entire recording and also transfer knowledge of these sound features to other recordings.

Summary

This chapter outlines an ear training method based on the source-destination audio editing technique. Because of the critical listening required to perform accurate audio editing, the process of finding and matching edit points can serve as an effective form of ear training. With the interactive software exercise module, the goal is to practice matching the length of one sound excerpt to a reference excerpt. By focusing on the timbre and amplitude envelope of the final milliseconds of the clip, the end point can be determined based on the nature of any transients or length of sustained signals. By not including verbal or meaningful numeric descriptors, the exercise is focused solely on the perceived audio signal and on matching the end point of the audio signals.

7

ANALYSIS OF SOUND

CHAPTER OUTLINE

Audio Production and Critical Listening. DOI: 10.1016/B978-0-240-81295-3.00007-1

After focusing on specific attributes of recorded sound, we are now ready to explore a broader perspective of sound quality and music production. Experience practicing with each of the software modules and specific types of processing described in the previous chapters prepares us to focus on these sonic features within a wider context of recorded and acoustic sound.

A sound recording is an interpretation and specific representation of a musical performance. Listening to a recording is different from attending a live performance, even for recordings with little signal processing. A sound recording can offer an experience that is more focused and clearer than a live performance, while also creating a sense of space. It is a paradoxical perspective to hear musicians with a high degree of clarity yet at the same time have the experience of listening from a more distant location because of the level of reverberant energy. Furthermore, a recording engineer and producer often make adjustments in level and processing over the course of a piece of music that highlight the most important aspects of a piece and guide a listener to a specific musical experience.

Each recording has something unique to tell in terms of its timbral, spatial, and dynamic qualities. It is important to listen to a wide variety of recordings from many different musical genres and to examine the production choices made for each recording. An engineer can familiarize herself with recording and mixing aesthetics for different genres of music that can inform her own work. When it comes time to make a recording, an engineer can rely on internal references for sound quality and mix balance to help guide a project. For each recording that seems interesting from a sound quality and production point of view, make a note of the production personnel credits, including producer, recording engineer, mixing engineer, and mastering engineer. With digitally distributed recordings, the production credits are not always listed with the audio but can be referenced through various websites such as www.allmusic.com. Finding additional recordings from engineers and producers previously referenced can help in the process of characterizing various production styles and techniques.

7.1 Analysis of Sound from Electroacoustic Sources

In the development of critical listening skills, it is necessary to examine, explore, and analyze sound recordings to help in understanding the sonic signatures of a particular artist, producer, or engineer. Through the analysis process it is possible to learn to identify what aspects of their recordings make them particularly successful from a timbral, spatial, and dynamic point of view.

The sound quality, technical fidelity, and sonic characteristics of a recording have a significant impact on how clearly the musical meaning and intentions of a recording are communicated to listeners. Components of a stereo image can be deconstructed to learn more about the use of reverberation and delays, panning, layering and balancing, dynamics processing, and equalization.

At its most basic level, the sound mixing process essentially involves gain control and level changes over time. Whether those changes are full-band or frequency selective, static or time varying, manual or through a compressor, the basic building block of sound mixing is control of sound level or amplitude. Single instruments or even single notes may be brought up or down in level to emphasize musical meaning.

In the critical listening and analysis process, there are numerous layers of deconstruction, from general, overall characteristics of a full mix to specific details of each sound source. At a much deeper level in the analysis of a recording, an engineer who is more advanced in critical listening skills can begin to make guesses about specific models of equipment used during recording and mixing, based on the timbres and amplitude envelopes of components in a sound image.

A stereo image produced by a pair of loudspeakers can be analyzed in terms of features that range from completely obvious to nearly imperceptible. A goal of ear training, as a type of perceptual learning, is to develop the ability to identify and differentiate features of a reproduced sound image, especially those that may not have been apparent before engaging in training exercises. We will now consider some of the specific characteristics of a stereo or surround image that are important to analyze. The list includes parameters outlined in the European Broadcasting Union Technical Document 3286 titled "Assessment Methods for the Subjective Evaluation of the Quality of Sound Programme Material—Music" (European Broadcasting Union [EBU], 1997):

- Overall bandwidth
- Spectral balance
- Auditory image
- Spatial impression, reverberation, and time-based effects
- Dynamic range, changes in level or gain, artifacts from dynamics processing (compressors/expanders)
- Noise and distortion
- Balance of elements within a mix

7.1.1 Overall Bandwidth

Overall bandwidth refers to the frequency content and how far it extends to the lowest and highest frequencies of the audio spectrum. In this part of the analysis, the goal is to determine if a recording extends from 20 Hz to 20 kHz, or if it is band-limited in some way. FM radio extends only up to about 15 kHz and the bandwidth of standard telephone communication ranges from about 300 to 3000 Hz. A recording may be limited by its recording medium, a sound system can be limited by its electronic components, and a digital signal may be down-sampled to a narrower bandwidth to save data transmission. The effect of narrowing a bandwidth can be heard through the use of high- and low- pass filters.

In making a judgment about high frequency extension, the highest overtones present in recording need to be considered. The highest fundamental pitches in music do not go much above about 4000 Hz, but overtones from cymbals and brass instruments easily reach 20,000 Hz. An engineer's choice of recording equipment or filters may intentionally reduce the bandwidth of a sound, which differentiates the bandwidth of the acoustic and recorded sound of an instrument.

7.1.2 Spectral Balance

As we saw in Chapter 2, spectral balance refers to the relative level of frequency bands across the entire audio spectrum. In its simplest analysis it can describe the balance of high frequencies to low frequencies, but it is possible to be more precise and identify specific frequency resonances and antiresonances. The power spectrum of an audio signal, which can help visualize a signal's spectral balance, can be measured in a number of ways. The most common calculation of power spectrum is probably through the fast Fourier transform (FFT), which specifies the frequency content of a signal and the relative amplitudes of frequency bands. The spectral balance of pink noise is flat when averaged over a period of time and graphed on a logarithmic frequency scale. Pink noise is perceived as having equal energy across the entire frequency range and therefore as having a flat spectral balance.

Through the subjective analysis of spectral balance, listen holistically to a recording. Where the possible combination and number of frequency resonances was simplified in Chapter 2, the analysis is now open to any frequency or combination of frequencies. Taking a broader view of a recording, the following questions are addressed:

- Are there specific frequency bands that are more prominent or deficient than others?
- Can we identify resonances by their approximate frequency in Hertz?
- Are there specific musical notes that are more prominent than others?

Frequency resonances in recordings can occur because of the deliberate use of equalization, microphone placement around an instrument being recorded, or specific characteristics of an instrument, such as the tuning of a drumhead. The location and angle of orientation of a microphone will have a significant effect on the spectral balance of the recorded sound produced by an instrument. Because musical instruments typically have sound radiation patterns that vary with frequency, a microphone position relative to an instrument is critical in this regard. (For more information about sound radiation patterns of musical instruments, see Dickreiter's book titled *Tonmeister Technology: Recording Environments, Sound Sources, and Microphone Techniques* [1989].) Furthermore, depending on the nature and size of a recording space, resonant modes may be present and microphones may pick up these modes. Resonant modes may amplify certain specific frequencies produced by the musical instruments. All of these factors contribute to the spectral balance of a recording or sound reproducing system and may have a cumulative effect if resonances from different microphones occur in the same frequency regions.

7.1.3 Auditory Image

An auditory image, as Woszczyk (1993) has defined it, is "a mental model of the external world which is constructed by the listener from auditory information (p. 198)." Listeners can localize sound images that occur from combinations of audio signals emanating from pairs or arrays of loudspeakers. The auditory impression of sounds located at various locations between two speakers is referred to as a *stereo image*. Despite having only two physical sound sources in the case of stereo, it is possible to create phantom images of sources in locations between the actual loudspeaker locations, where no physical source exists.

Use of a complete stereo image—spanning the full range from left to right—is an important and sometimes overlooked aspect of production. Careful listening to recordings can illustrate a variety of panning and stereo image treatments. The illusion of a stereo image is created by controlling interchannel amplitude differences through panning and interchannel time differences through time delay. Interchannel differences do not correspond to inter*aural* differences when reproduced over loudspeakers because sound from both loudspeakers reaches both ears. Stereo microphone techniques can provide yet another method to control interchannel amplitude and time differences because of microphone polar patterns and physical spacing between microphones.

In the study of music production and mixing techniques, a number of conventions in panning sounds within the stereo image are found among various genres of music. For instance, pop and rock generally emphasize the central part of the stereo image, because kick drum, snare drum, bass, and vocals are typically panned to the center. Guitar and keyboard parts are sometimes panned to the side, but overall there is significant energy originating from the center. A look at a correlation meter would confirm what is heard as well, and a recording with a strong center component will give a reading near 1 on a correlation meter. Likewise, if the polarity of one channel is reversed and the left and right channels are added together, a mix with a dominant center image will have significant cancellation of the audio signal. Any audio signal components that are equally present in the left and right channels (i.e., panned center or monophonic) will have destructive cancellation when the two channels are subtracted.

Panning and placement of sounds in a stereo image have a definite effect on how clearly listeners can hear individual sounds in a mix. The phenomenon of masking, where one sound obscures another, should also be considered with panning. Panning sounds apart will result in greater clarity, especially if they occupy similar musical registers or contain similar frequency content. The mix and musical balance, and therefore the musical meaning and message, of a recording are directly affected by the panning of instruments, and the appropriate use of panning can give an engineer more flexibility for level adjustments.

While listening to stereo image width and the spread of an image from one side to the other, the following questions guide the exploration and analysis:

- Taken as a whole, does a stereo image have a balanced spread from left to right with all points between the loudspeakers being equally represented or are there locations where it seems like an image is lacking?
- How wide or monophonic is the image?
- What are the locations and widths of individual sound sources in a recording?
- Are their locations stable and definite or ambiguous?
- How easily can the locations of sound sources be pinpointed within a stereo image?
- Does the sound image appear to have the correct and appropriate spatial distribution of sound sources?

By considering these types of questions for each sound recording encountered, a stronger sense can be developed for the kinds of panning and stereo images created by professional engineers and producers.

7.1.4 Spatial Impression, Reverberation, and Time-Based Effects

The spatial impression of a recording is critical for conveying emotion and drama in music. Reverberation and echo help set the scene in which a musical performance or theatrical action takes place. Listeners can be transported mentally to the space in which music exists through the strong influence of early reflections and reverberation that envelops music in a sound recording. Whether a real acoustic space is captured in a recording or artificial reverberation is added to mimic a real space, spatial attributes convey a general impression about of the size of a space. A long reverberation time can create the sense of being in a larger acoustic space, whereas a short reverberation decay time or a low level of reverberation can convey the feeling of a more intimate, smaller space.

The analysis of spatial impression can be broken down into the following subareas:

- Apparent room size:
  - How big is the room?
  - Is there more than one type of reverberation present in a recording?
  - Is the reverberation real or artificial?
  - What is the approximate reverberation time?
  - Are there any echoes or long delays in the reverberation and early reflections?
- Depth perspective: Are sounds placed up front clearly distinguished from those in the background?
- What is the spectral balance of the reverberation?
- What is the direct/reverberant ratio?
- Are there any strong echoes or delays?
- Is there any apparent time-based effect such as chorus or flanging?

Classical music recordings can give listeners the opportunity to familiarize themselves with reverberation from a real acoustic space. Often orchestras and artists with higher recording budgets will record in concert halls and churches with acoustics that are deemed to be very conducive to music performance. The depth and sense of space that can be created with appropriate pickup of a real acoustic space are generally difficult to mimic with artificial reverberation.

Adding artificial reverberation to dry sounds is not the same as recording instruments in a live acoustic space from the start. If dry sound is recorded in an acoustically dead space with close microphones, then microphones are not picking up sound that is radiating away from the microphones. Sound that is radiated from the back of an instrument will not likely be picked up in a dry studio environment. So even when artificial reverberation of the highest quality is added, it will not sound the same as an instrument recorded in a live acoustic space with close and room microphones.

7.1.5 Dynamic Range and Changes in Level

Dynamic range can be critical to a music recording, and different styles of music will require different dynamic ranges. There may be broad fluctuations in sound level over the course of a musical piece, as a dynamic level rises to fortissimo and falls to pianissimo. Likewise the micro-dynamics of a signal can be examined, the analysis of which is usually aided by the use of a level meter such as a peak program meter (PPM) or digital meter. For pop and rock recordings, usually the dynamic range from a level point of view is fairly static, but we may hear (and see on a meter) small fluctuations that occur on strong beats and in between those beats. A meter may fluctuate more than 20 dB for some recordings or as few as 2 to 3 dB for others. Fluctuations of 20 dB represent a wider dynamic range than smaller fluctuations, and they usually indicate that a recording has been compressed less. Because the human auditory system responds primarily to average levels rather than peak levels in the judgment of loudness, a recording with smaller amplitude fluctuations will sound louder than one with larger fluctuations, even if the two have the same peak amplitude.

In this part of the analysis, listen for changes in level of individual instruments and of an overall stereo mix. Changes in level may be the result of manual gain changes or automated, signal-dependent gain reduction produced by a compressor or expander. Dynamic level changes can help magnify musical intentions and enhance the listening experience. A downside to a wide dynamic range is that the quieter sections are partially inaudible and thus detracting from any musical impact intended by an artist.

7.1.6 Noise and Distortion

Many different types of noise can disrupt or degrade an audio signal in one way or another and can come in different forms such as 50 or 60 Hz buzz or hum, low-frequency thumps from a microphone or stand being bumped, external noises such as car horns or airplanes, clicks and pops from inaccurate digital synchronization, and dropouts (very short periods of silence) resulting from defective recording media. Generally the goal is to avoid any accidental instances of noise, unless of course, they suit a deliberate artistic effect.

Unless intentionally distorting a sound, engineers try to avoid clipping any of the stages in a signal chain. So it is important to recognize when it is occurring and reduce a signal's level appropriately. Sometimes it is unavoidable or it slips by those involved and is present in a finished recording.

7.1.7 Balance of the Components within a Mix

Finally, in the analysis of recorded sound, consider the mix or the balance of the elements within a recording. The relative balance of instruments can have a highly significant influence on musical meaning, impact, and focus of a recording. The amplitude of one element within the context of a mix can also have an effect on the perception of other elements within the mix.

Think about questions such as the following:

- Are the amplitude levels of the instruments balanced appropriately for the style of music?
- Is there an instrument that is too loud or another that is too quiet?

The entire perceived sound image can be analyzed as a whole. Likewise, less significant features of a sound image may be analyzed as well and can be considered as a subgroup. Some of these subfeatures might include the following:

- Specific features of each component, musical voice, or instrument such as the temporal nature or spatial location

of amplitude envelope components (e.g., attack, decay, sustain, and release)
- Definition and clarity of elements within a sound image
- Width and spatial extent of individual elements

Often, for an untrained listener, specific features of reproduced audio may not be obvious or immediately recognizable. A trained listener on the other hand will likely be able to identify and distinguish specific features of reproduced audio that are not apparent to an untrained listener. There is such an example in the world of perceptual encoder algorithm development, which has required the use of expert trained listeners to identify shortcomings in the processing. Artifacts and distortion produced during perceptual encoding are not necessarily immediately apparent until critical listeners, who are testing encoding software, learn what to listen for. Once a listener can identify audio artifacts, it can become difficult *not* to hear them.

Distinct from listening to music at a live concert, music recordings (audio-only, as opposed to those accompanied by video) require listeners to rely entirely on their sense of hearing. There is no visual information to help follow a musical soundtrack, unlike a live performance where visual information helps to fill in details that may not be as obvious in the auditory domain. As a result, recording engineers sometimes exaggerate certain sonic features of a sound recording, through level control, dynamic range processing, equalization, and reverberation, to help engage a listener.

7.2 Analysis Examples

In this section we will do a survey of some recordings, highlighting timbral, dynamic, spatial, and mixing choices that are apparent from listening. Any of these tracks would be appropriate for practicing with the EQ software module, auditioning loudspeakers and headphones, and doing graphical analysis (see Section 7.3).

7.2.1 Sheryl Crow: "Strong Enough"

Crow, Sheryl. (1993). *Tuesday Night Music Club*. A&M Records. Produced by Bill Bottrell.

The third track from Sheryl Crow's *Tuesday Night Music Club* is fascinating in its use of numerous layers of sounds that are arranged and mixed together to form a musically and timbrally interesting track. The instrumental parts complement each other and are well balanced. Numerous auditions of the track are required to identify all the sounds that are present.

The piece starts with a synthesizer pad followed by two acoustic guitars panned left and right. The guitar sound is not as crisp sounding as might be imagined from an acoustic guitar. In this recording, the high frequencies of these guitars have been rolled off somewhat, perhaps because the strings are old and some signal from an acoustic guitar pickup is mixed in.

Crow's lead vocal enters with a dry yet intense sound. There is very little reverb on the voice, and the timbre is fairly bright. A crisp, clear 12-string comes in contrasting with the dull sound of the other two guitars. Fretless electric bass enters to round out the lower pitches. Hand percussion is panned left and right to fill out the spatial component of the stereo image.

The chorus features a fairly dry ride cymbal and a high, flutey Hammond B3 sound fairly low in the mix. After the chorus a pedal steel enters and then fades away before the next verse. The bridge features bright and clear, strumming mandolins that are panned left and right. Backing vocals, panned left and right, echo Crow's lead vocal.

The instrumentation and unconventional layering of contrasting sounds makes this recording interesting from a subjective analysis point of view. The arrangement of the piece results in various types of instruments coming and going to emphasize each section of the music. Despite the coming and going of instruments and the number of layers present, the music sounds clear and coherent.

7.2.2 Peter Gabriel: "In Your Eyes"

Gabriel, Peter. (1986). *So.* Produced by Daniel Lanois and Peter Gabriel. Engineered by Kevin Killen and Daniel Lanois. The David Geffen Company.

This track by Peter Gabriel is a study in successful layering of sounds that create a full mix timbrally, dynamically, and spatially. The music starts with chorused piano sound, synthesizer pad, and percussion. Bass and drums enter soon after, followed by Gabriel's lead vocal.

There is an immediate sense of space on the first note of the track. There is no obvious reverberation decay in the beginning, yet the combination of all of the sounds each with its own sense of space creates an open feeling. Reverberation decay is more audible after the chorus when the percussion and synthesizers vamp for a few measures.

Despite multiple layers of percussion such as talking drum and triangle, along with the full rhythm section, the mix is pleasingly full yet remains uncluttered. The various percussion parts and drum kit occupy a wide area in the stereo image, helping to create a space in which the lead vocal sits.

The vocal timbre has a warm yet slightly gritty sound. It is completely supported by the variety of drums, bass, percussion, and synthesizers through the piece. Senegalese singer Youssou N'Dour performs a solo at the end of the piece, which is layered with other vocals that are panned out to the sides. The bass line is punchy and articulate, sounding as though it was compressed fairly heavily, and it contributes significantly to the rhythmic foundation of the piece.

Distortion is present in a few sounds, starting with the slightly crunchy drum hit on the downbeat of the piece. Other sounds are slightly distorted in places and compression effects are audible. This is certainly not the cleanest recording to be found, yet the distortion and compression artifacts work to add life and excitement to the recording.

Overall this recording demonstrates a fascinating use of many layers of sound, including acoustic percussion and electronic synthesizers, which create the sense of a large open space in which a musical story is told.

7.2.3 Lyle Lovett: "Church"

Lovett, Lyle. (1992). *Joshua Judges Ruth.* Produced by George Massenburg, Billy Williams, and Lyle Lovett. Recorded by George Massenburg and Nathan Kunkel. Curb Music Company/MCA Records.

Lyle Lovett's recording of "Church" represents contrasting perspectives. The track begins with piano giving a gospel choir a starting note, which they hum. Lovett's lead vocal enters immediately with handclapping from the choir on beats two and four. The piano, bass, and drums begin some sparse accompanying of the voice and gradually build to more prominent parts. One thing that is immediately striking in this recording is the clarity of each sound. The timbres of instruments and voices represent evenly balanced spectra, coming forth from the mix as natural sounding.

Lovett's vocal is up front with very little reverberation, and its level in the mix is consistent from start to finish. The drums have a crisp attack with just the right amount of resonance. Each drum hit pops out from the mix with toms panned across the stereo image. The cymbals are crystal clear and add sparkle to the top end of the recording.

The choir in this recording accompanies Lovett and responds to his singing. Interestingly, the choir sounds like it is set in a small country church, where the reverberation is especially highlighted by hand claps. The choir and associated handclaps are panned widely across the stereo image. As choir members take short solos, their individual voices come forward and are particularly drier than they are when with the choir.

The lead vocals and rhythm section are presented in a fairly dry, up front way, and this contrasts with the choir, which is clearly in a more reverberant space or at least more distant.

Levels and dynamic range of each instrument are properly adjusted, presumably through some combination of compression and manual fader control. Each component of the mix is audible and none of the sounds is obscured.

Noises and distortion are completely nonexistent in this recording, and obviously great care has been taken to remove or prevent any extraneous noise. There is also no evidence of clipping, and each sound is clean.

This recording has become a classic in terms of sound quality and has also been mixed in surround as a separate release.

7.2.4 Sarah McLachlan: "Lost"

McLachlan, Sarah. (1991). *Solace*. Produced and recorded by Pierre Marchand. Nettwerk/Arista Records, Bertelsmann Music Group.

This track starts with a somewhat reverberant yet clear acoustic guitar and focused, dry brushes on a snare drum. A somewhat airy lead vocal enters with a large space around it. The reverb creating the space around the voice is fairly low in level, but the decay time is probably in the 2-second range. The reverberation blends well with the voice and seems to be appropriate for the character of the piece. The timbre of the voice is clear and balanced spectrally. Mixing and compression of the voice has made its level consistently forward of the ensemble.

Mandolin and 12-string guitar panned slightly left and right enter after the first verse along with electric bass and reverberant pedal steel. The bass plays a few pitches below the standard low notes of a bass, creating an enveloping sound that supports the rest of the mix. Backing vocals are panned slightly left and right and are placed a bit farther back in the mix than the lead vocal. Synthesizer pads, backing vocals, and delayed guitar transform the mix into a dreamy texture for a verse and then fade out for a return of the mandolin and 12-string guitar.

The timbres in this track are clear yet not harsh. There is an overall softness to the timbres and the low frequencies—mostly from the bass—provide a solid foundation for the mix. (Interestingly, some sounds on other tracks on this album are slightly harsh sounding.) The lead vocal is the most prominent sound in the mix with backing vocals mixed slightly lower than the lead vocal. Guitars, mandolin, and bass are the next most prominent sound in the mix. Drums are almost completely gone after the intro but return at the end. The drummer elevates the energy of the final chorus by playing tom and snare hits. The drums are mixed fairly low but are still audible as a rhythmic texture and the snare drum has the snares disengaged.

With the round, smooth, full sound of the bass, this recording is useful for hearing the low-frequency response of loudspeakers and headphones. There is not much attack on the bass to identify articulation, but its sound suits the music comfortably. With such a prominent and balanced vocal, the recording can also serve to help identify any mid-frequency resonances or antiresonances in a sound reproduction system.

7.2.5 Jon Randall: "In the Country"

Randall, Jon. (2005). *Walking Among the Living*. Produced by George Massenburg and Jon Randall. Recorded by George Massenburg and David Robinson. Epic/Sony BMG Music Entertainment.

The fullness and clarity of this track are present from the first note. Acoustic guitar and mandolin start the introduction followed by Randall's lead vocal. The rhythm section enters in the second verse, which extends the bandwidth with cymbals in the high-frequency range and kick drum in the low-frequency range. Various musical colors, such as dobro, fiddle, Wurlitzer, and mandolin, rise to the forefront for short musical features and then fade to the background. It seems apparent that great care was taken to create a continually evolving mix that features musically important phrases.

The timbres in this track sound naturally clear and completely balanced spectrally. The voice is consistently present above the instruments, with a subtle sense of reverberation to create a space around it. The drums are not as prominent as they are on the Lyle Lovett recording discussed earlier, and they are a little understated. The cymbals are present and clear, but they do not overpower other sounds. The bass is smooth and full, with enough articulation for its part. The fiddle, mandolin, and guitar sounds are all full bodied, crisp, and warm. The high harmonics of the strummed mandolin and guitars blend with the cymbals' harmonics in the upper frequency range. Further to the timbral integrity of the track, there is no evidence of any noise or distortion.

The stereo image is used to its full extent with mandolins, guitars, and drum kit panned wide. The balance on this recording is impeccable and makes use of musically appropriate spatial treatment (reverberation and panning), dynamics processing, and equalization.

7.3 Graphical Analysis of Sound

In research on the perception of sound images produced by car audio systems, researchers have used graphical techniques to elicit listeners' perceptions of the location and dimensions of sound images (Ford et al., 2002, 2003; Mason et al., 2000). Work done by Usher and Woszczyk (2003) and Usher (2004) has sought to visualize the placement, depth, and width of sound images within a multichannel reproduction environment, to better understand listeners' perceptions of sound source locations in an automotive sound reproduction environment. In the experiments, listeners were asked to draw sound sources using elliptical shapes on a computer graphical interface.

By translating what is heard to a visual, two-dimensional diagram, a level of analysis can be achieved distinct from verbal descriptions. Although there is no clear-cut method for visually illustrating an auditory perception, the exercise of doing so is very useful for sonic analysis and exploration.

Using a template such as in Figure 7.1, draw what is heard coming from a sound system. The listening location relative to a sound system will have a direct effect on the

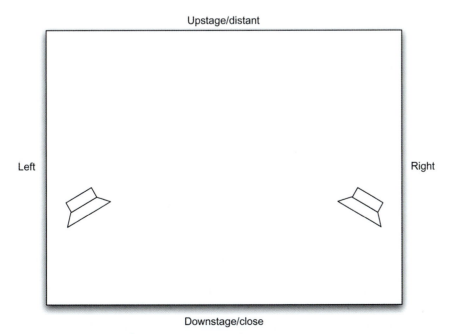

Figure 7.1 The reader is encouraged to use the template shown here as a guide for the graphical analysis of a sound image, to visualize the perceived locations of sound images within a sound recording.

localization of phantom images. Section 1.3.2 illustrates the ideal listening location for stereo sound reproduction that will give accurate phantom image locations.

The images that are drawn on the template should not resemble actual musical instrument shapes but should be analogous to the sound images that are perceived from the loudspeakers. For example, the stereo image of a solo piano recording will be much different from the image of a piano playing with an ensemble, and their corresponding visual images would also look significantly different.

Drawings of stereo images should be labeled to indicate how the visual forms correspond to the perceived aural images. Without labels they may appear too abstract to be understood, but when considered in relation to their respective sound recordings they can help a listener map out a sound image.

Graphical analysis allows the focus to be on the location, width, depth, and spread of sound sources in a sound image. A visual representation of a sound image should include not only direct sound from each sound source but also any spatial effects such as reflections and reverberation present in a recording.

7.4 Multichannel Audio

This section will focus on the most common multichannel reproduction format with 5.1 channels. Multichannel audio generally allows the most realistic reproduction of an enveloping sound field, especially for recordings of purely acoustic music in a concert hall setting; this type of recording can leave listeners with the impression that they are seated in a hall, completely enveloped by sound.

Conversely, multichannel audio also offers the most unrealistic reproduction of audio because it allows an engineer to position sound sources around a listener. Typically there are no musicians placed behind audience members at a concert, other than antiphonal organ, brass or choir, but multichannel audio reproduction allows a sound mixer to place direct sound sources to the rear of the listening position. Certainly multichannel audio has many advantages over two-channel stereo, but there are still challenges to be considered and opportunities for critical listening to help with these challenges.

Although there are loudspeakers in front and behind, in the ITU-R BS.775-1 (ITU-R, 1994) recommendation (see Fig. 1.3) there exists a fairly wide space between the front loudspeaker (30°) and the nearest surround loudspeaker (at 110° to 120°). The wide space between front and rear loudspeakers makes lateral sound images difficult to produce, at least with any stability and locational accuracy.

7.4.1 The Center Channel

A distinctive feature of the 5.1 reproduction environment is the presence of a center speaker situated at 0° between the left and right channels. The advantage of a center channel is that it can help solidify and stabilize sound images that are panned to the center. Phantom images in the center of a conventional stereo loudspeaker setup appear to come from the center only when the listener is seated in the ideal listening location, equidistant from the loudspeakers. When a listener moves to one side, a central phantom image appears to move to the same side. Because a listener is no longer equidistant from the two loudspeakers, sound arrives at the listener from the closest speaker first and will be localized at that speaker because of the law of first arriving wavefront, also known as the *precedence effect* or *Haas effect*.

Soloing the center speaker of a surround mix helps give an idea of what a mix engineer sent to the center channel. When listening to the center channel and exploring how it is integrated with the left and right channels, ask questions such as the following:

- Does the presence or absence of the center channel make a significant difference to the front image?
- Are lead instruments or vocals the only sounds in the center channel?
- Are any drums or components of the drum kit panned to the center channel?
- Is the bass present in the center channel?

If a recording has prominent lead vocals and they are panned only to the center channel, then it is likely that some of the reverberation, echo, and early reflections are panned to other channels. In such a mix, muting the center channel can make it easier to hear the reverberation without any direct sound.

Sometimes phantom images produced by the left and right channels are reinforced by the center image or channel. Duplicating a center phantom image in the center speaker, can make the central image more stable and solid. Often the signal that is sent to the left and right channels may be delayed or altered in some way so that it is not an exact copy of the center channel. With all three channels producing exactly the same audio signal, the listener can experience comb filtering with changes in head location as the signals from three different locations combine at the ears (Martin, 2005).

The spatial quality of a phantom image produced between the left and right channels is markedly different from the solid image of the center channel reproducing exactly the same audio signal on its own. A phantom image between the left and right loudspeakers may still be preferred by some despite its shortcomings, such as phantom image movement corresponding to listener location. A phantom image produced by two loudspeakers will generally be wider and more full sounding than a single center loudspeaker producing the same sound, which may be perceived as narrower and more constricted.

It is important to compare different channels of a multichannel recording and start to form an internal reference for various aspects of a multichannel sound image. By making these comparisons and doing close, careful listening we can form solid impressions of what kinds of sounds are possible from various loudspeakers in a surround environment.

7.4.2 The Surround Channels

In the analysis of surround recordings, it is useful to focus on how well a recording in 5.1-channel surround achieves a smooth spread from front to rear and if a side image exists. Side images are difficult to produce without an actual loudspeaker positioned on the side because of the nature of binaural hearing, which is far more accurate at localizing sounds originating from the front.

Localize various elements in a mix and examine the placement of sounds around the listening area by considering some questions such as:

- How are different elements in the mix panned?
- Do they have precise locations, or is it difficult to determine the exact location because a sound seems to come from many locations at once?
- What is the nature of the reverberation and where is it panned?
- Are there different levels of reverberation and delay?

In surround playback systems, the rear channels are widely spaced. The wide spacing, coupled with binaural hearing that has less spatial acuity in the rear, makes it challenging to create a cohesive, evenly spread rear image. It is important to listen to just the surround channels. When listening to the entire mix, the rear channels may not be as easy to hear because of the auditory system's predisposition to sound arriving from the front.

7.4.3 Exercise: Comparing Stereo to Surround

Comparing a stereo and surround mix of the same musical recording can be enlightening. Many details can be heard in a surround mix that are not as audible or are missing from a stereo mix. Surround reproduction systems allow an engineer to place sound sources at many different locations around a listening area. Because of the spatial separation of sound sources, there is less masking in a surround mix. Listening to a surround mix and then switching back to its corresponding stereo mix can help highlight elements of a stereo mix that were not audible before.

7.4.4 Exercise: Comparing Original and Remastered Versions

A number of recordings have been remastered and rereleased several years after their original release. Remastering an album usually involves returning to its original stereo mix and applying new equalization, dynamics processing, level adjustments, mid-side processing, and possibly reverberation. Comparing an original release of an album to a remastered version is a useful exercise that can help highlight timbral, dynamic, and spatial characteristics typically altered by a mastering engineer.

7.5 High Sampling Rates

There have been a number of heated debates about the advantages or benefits of high sampling rates in digital audio. The compact disc digital audio format specifies a sampling rate of 44,100 Hz and a bit depth of 16 bits per sample, according to the Red Book CD standard. As recording technology has evolved, it has allowed recording and distribution of audio to listeners at much higher sampling rates. There is no question that bit depths greater than 16 bits per sample improve audio quality, and engineers typically record with at least 24 bits per sample. As an exercise, compare a 24-bit recording to a dithered down 16-bit version of the same recording and note audible differences.

Sampling rate determines the highest frequency that can be recorded and therefore the bandwidth of a recording. Sampling theorem states that the highest frequency we can record is equal to half the sampling frequency. Higher sampling rates allow a wider bandwidth for recording.

Though the difference between a high sampling rate (96 kHz or 192 kHz) and 44.1 kHz sampling rate is subtle, and it may be difficult to hear any difference, comparing high sampling rate with CD-quality audio can be helpful in fine-tuning listening skills. As one progresses to perceiving finer audible differences between sounds, it can be useful to compare sound recorded at different sampling rates. Some engineers report that a recording made at 96 kHz and down-sampled to 44.1 kHz sounds better than a recording that originates at 44.1 kHz.

A 2.8224 MHz sampling rate recording from a Super Audio CD (SACD) may offer a greater difference than 96 kHz or 192 kHz when compared to CD-quality audio. One of the differences has to do with improved spatial clarity. The panning of instruments and sound sources within a stereo or surround image can be more clearly defined, source locations are more precise, and reverberation decay is generally smoother.

With any of these comparisons, it is easier to hear differences when the audio is reproduced over high-quality loudspeakers or headphones. Lower-quality reproduction devices do not allow full enjoyment of the benefits of high sampling rates. High-quality reproduction systems do not always have to be expensive, especially in consumer systems.

7.6 Exercise: Comparing Loudspeakers and Headphones

Each particular model of loudspeaker or headphone has a unique sound. Frequency response, power response, distortion characteristics, and other specifications all contribute to the sound that an engineer hears and thus influence decisions during recording and mixing sessions.

For this exercise, do the following:

- Choose either two different pairs of speakers, two different headphones, or a pair of loudspeakers and a pair of headphones.
- Choose several familiar music recordings.
- Document the make/model of the loudspeakers/headphones and listening environment.
- Compare the sound quality of the two different sound reproduction devices.
- Describe the audible differences with comments on the following aspects and features of the sound field:
  - Timbral quality—Describe differences in frequency response and spectral balance.
    - Is one model deficient in a specific frequency band?
    - Is one model particularly resonant in a certain frequency band?
  - Spatial characteristics—How does the reverberation sound?
    - Does one model make the reverberation more prominent than the other?
    - Is the spatial layout of the stereo image the same in both?
    - Is the clarity of sound source locations the same in both? That is, can sound sources be localized in the stereo image equally well in both models?

- – If comparing headphones to loudspeakers, can we describe differences in those components of the image that are panned center?
- – How do the central images compare in terms of their location front/back and their width?
- • Overall clarity of the sound image:
  - – Which one is more defined?
  - – Can details be heard in one that are less audible or inaudible in the other?
- • Preference—Which one is preferred overall?
- • Overall differences—Describe any differences beyond the list presented here.
- • Sound files—It is best to use only linear PCM files (AIFF or WAV) that have not been converted from MP3 or AAC.

Each sound reproducing device and environment has a direct effect on the quality and character of the sound heard, and it is important for an engineer to know his sound reproduction system (the speaker/room combination) and have a couple of reference recordings that he knows well. Reference recordings do not have to be pristine, perfect recordings as long as they are familiar.

7.7 Exercise: Sound Enhancers on Media Players

Many software media players used for playing audio on a computer offer so-called sound enhancement controls. This type of control is often turned on by default in media players such as iTunes, and it offers another opportunity for critical listening. It can be informative to compare the audio quality with the sound enhancement on and off and try to determine by ear how the algorithm is affecting the sound. The processing that it employs may improve the sound of some recordings but degrade the sound of others.

Consider how a sound enhancer affects the stereo image and if the overall image width is affected or if panning and location of sound sources are altered in any way:

- • Is the reverberation level affected?
- • The timbre will likely be altered in some way. Try to identify as precisely as possible how the timbre is

changed. Identify if any equalization has been added and what specific frequencies have been altered.

- Is there any dynamic range processing occurring? Are there artifacts of compression present or does the enhanced version sound louder?

The sound enhancement setting on media players may or may not be altering audio in a desirable way, but it certainly offers a critical listening exercise in determining the differences in audio characteristics.

7.8 Analysis of Sound from Acoustic Sources

Live acoustic music performances can be instructive and enlightening in the development of critical listening skills. The majority of the music heard is through electroacoustic transducers of some sort (loudspeakers or headphones), and it can be easy to lose sight of what an instrument sounds like acoustically, as it projects sound into all directions in a room or hall. At least one manufacturer of consumer audio systems encourages its research and development staff to attend concerts of acoustic music. This practice is incredibly important for developing a point of reference for tuning loudspeakers. The act of listening to sound quality, timbre, spatial characteristics, and dynamic range during a live music concert can fine-tune skills for technical listening over loudspeakers.

It may seem counterintuitive to use such acoustic music performances for training in a field that relies on sound reproduction technology, but the sound radiation patterns of musical instruments are different from those of loudspeakers, and it is important to recalibrate the auditory system by listening actively to acoustic music. When attending concerts of jazz, classical, contemporary acoustic music, or folk music, the result of each instrument's natural sound radiation patterns into the room can be heard. Sound emanates from each instrument into the room, theater, or hall and mixes with that from other instruments and voices.

Seated in the audience at a concert of live music, focus on aspects of the sound that are often considered when balancing tracks in a recording. Just as the spatial layout (panning) and depth of a recording reproduced over loud-speakers can be analyzed, these aspects can also be examined in an acoustic setting. Begin by trying to localize the various members or sections of the ensemble that is performing. With eyes closed it may be easier to focus on the aural sensation and ignore what the sense of sight is reporting. Attempt to localize instruments on a stage and think about the overall sound in terms of a "stereo image"—as if two loudspeakers were producing the sound and phantom images are heard between the speakers. The localization of sound sources may not be the same for all seats in the house and may be influenced by early reflections from side walls in the performance space. When comparing music being reproduced over a pair of loudspeakers to that being performed in a live acoustic space, the sound image perceived is going to be significantly different in terms of timbre, space, and dynamics. Some questions can guide the comparison:

- Does the live music sound wider overall or narrower than stereo loudspeakers?
- Is the direct to reverberant ratio consistent with what might be heard in a recording?
- How does the timbre compare to what is heard over loudspeakers? If it is different, describe the difference.
- How well are very quiet passages heard?
- How does the dynamic range compare?
- How does the sense of spaciousness and envelopment compare?

Audience members almost always sit much farther away from musical performers than microphones would typically be placed, and they are outside of the reverberation radius or critical distance. Therefore, the majority of the sound energy that they are hearing is indirect sound—reflections and reverberation—so it is therefore much more reverberant than what is heard on a recording. This level of reverberation would not likely be acceptable in a recording, but audience members find it enjoyable. Perhaps because music performers are visible in a live setting, the auditory system is more forgiving, or perhaps the visual cues help audience members engage with the music because they can see the movements of the performers in sync with the notes that are being played.

Ideally the reverberant field—the audience seating area—should be somewhat diffuse, meaning indirect sound should be heard coming equally from all directions. In a real concert hall or other music performance space, this may not be the case and it may be possible to localize the reverberation. If the reverberation is localizable, then focus on the width and spatial extent of it. Is it primarily located behind or does it also extend to the sides? Is it enveloping? Is there any reverberation coming from the front where the musicians are typically located?

Early reflections may also be discernible as a feature of any sound field. Although early reflections usually arrive at the listener within tens of milliseconds of a direct sound and are therefore imperceptible as discrete sounds, there are occasions when reflections can build up or become focused from a particular location and alter our perception of the location of a sound source. Any curved wall will tend to focus reflections, causing them to add together and therefore increase in amplitude to a level greater than the direct sound.

Early reflections from the side can help to broaden the perceived width of the sound image. Although these reflections may not be perceivable as discrete echoes, try to focus on the overall width. Focus also on how the direct sound blends and joins the sound coming from the sides and rear. Is the sound continuously enveloping all around, or are there breaks in the sound field, as there may be when listening to multichannel recordings?

Echoes, reflections, and reverberation are sometimes more audible when transient or percussive sounds are present. Sounds that have a sharp attack and short sustain and decay will allow indirect sound that comes immediately after it to be heard, because the direct sound will be silent and therefore will not mask the indirect sound.

Summary

The analysis of sound, whether purely acoustic or originating from loudspeakers, presents opportunities to deconstruct and uncover characteristics and features of a sound image. The more one listens to recordings and acoustic sounds with active engagement, the more sonic features one is able to pinpoint and focus on. With time and continued practice, the perception of auditory events opens up and one begins to notice sonic characteristics that were previously not audible. The more one uncovers through active listening, the deeper the enjoyment of sound can become, but it does take dedicated practice over time. Likewise, more focused and effective listening skills lead to improved efficiency and effectiveness in sound recording, production, composition, reinforcement, and product development. Technical ear training is critical for anyone involved in audio engineering and music production, and critical listening skills are well within the grasp of anyone who is willing to spend time being attentive to what he or she is hearing.

Here are some final words of advice: Listen to as many recordings as possible. Listen over a wide variety of headphones and loudspeaker systems. During each listening session, makes notes about what is heard. Find out who engineered the recordings that are most admired and find more recordings by the same engineers. Note the similarities and differences among various recordings by a given engineer, producer, or record label. Note the similarities and differences among various recordings by a given artist who has worked with a variety of engineers or producers.

The most difficult activity to engage in while working on any audio project is continuous active listening. The only way to know how to make decisions about what gear to use, where to place microphones, and how to set parameters is by listening intently to every sound that emanates from one's monitors and headphones. By actively listening at all times, one can gain essential information to best serve the musical vision of any audio project. In sound recording and production, the human auditory system is the final judge of quality and artistic vision.

BIBLIOGRAPHY

Barron, M. (1971). The subjective effects of first reflections in concert halls: The need for lateral reflections. *Journal of Sound and Vibration, 15*, 475–494.

Bech, S. (1992). Selection and training of subjects for listening tests on sound-reproducing equipment. *Journal of the Audio Engineering Society, 40*, 590–610.

Blesser, B., & Salter, L.-R. (2006). *Spaces speak, are you listening? Experiencing aural architecture*. Cambridge, MA: MIT Press.

Bradley, J. S., & Soulodre, G. A. (1995). Objective measures of listener envelopment. *Journal of the Acoustical Society of America, 98*(5), 2590–2597.

Brixen, E. B. (1993). Spectral ear training. *Proceedings of the 94th Convention of the Audio Engineering Society*, Preprint 3474, Berlin, Germany.

Case, A. U. (2007). *Sound FX: Unlocking the creative potential of recording studio effects*. Boston: Focal Press/Elsevier.

Corey, J. (2002). *An integrated system for dynamic control of auditory perspective in a multichannel sound field*. Ph.D. thesis. Montreal, Canada: McGill University, http://www-personal.umich.edu/~coreyja.

Corey, J. (2004). An ear training system for identifying parameters of artificial reverberation in multichannel audio. *Proceedings of the 117th Convention of the Audio Engineering Society*, Preprint 6262, San Francisco.

Corey, J. (2007). Beyond splicing: Technical ear training methods derived from digital audio editing techniques. *Proceedings of the 123rd Convention of the Audio Engineering Society*, Preprint 7212, New York.

Corey, J., Woszczyk, W., Martin, G., & Quesnel, R. (2001). An integrated multidimensional controller of auditory perspective in a multi-channel soundfield. *Proceedings of the 111th Convention of the Audio Engineering Society*, Preprint 5417, New York.

Dickreiter, M. (1989). *Tonmeister technology: Recording environments, sound sources, and microphone techniques*. New York: Temmer Enterprises.

European Broadcasting Union (EBU) (1997). Assessment methods for the subjective evaluation of the quality of sound programme material—Music. Technical Document 3286-E, European Broadcasting Union, Geneva, Switzerland.

Fletcher, H., & Munson, W. A. (1933). Loudness, its definition, measurement and calculation. *Journal of the Acoustical Society of America, 5*(2), 82–108.

Ford, N., Rumsey, F., & Nind, T. (2002). Subjective evaluation of perceived spatial differences in car audio systems using a graphical assessment language. *Proceedings of the 112th Convention of the Audio Engineering Society*, Preprint 5547, Munich, Germany.

Ford, N., Rumsey, F., & Nind, T. (2003). Evaluating spatial attributes of reproduced audio events using a graphical assessment language: Understanding differences in listener depictions. *Proceedings of the Audio Engineering Society 24th International Conference*, Banff, Canada.

Geddes, E., & Lee, L. (2003). Auditory perception of nonlinear distortion: Theory. *Proceedings of the 115th Convention of the Audio Engineering Society*, Preprint 5890, New York.

Gerzon, M. A. (1986). Stereo shuffling: New approach—old technique. *Studio Sound, 28*, 122–130.

Gerzon, M. A. (1994). Applications of blumlein shuffling to stereo microphone techniques. *Journal of the Audio Engineering Society, 42*(6), 435–453.

Getz, S., Gilberto, J., & Jobim, A. (1964). "Desafinado" from Getz/Gilberto. Verve Records.

Gibson, E. J. (1969). *Principles of perceptual learning and development.* New York: Appleton-Century-Crofts.

Giddings, P. (1990). *Audio systems design and installation.* Boston: Focal Press.

Howard, D., & Angus, J. A. S. (2006). *Acoustics and psychoacoustics* (3rd ed.). Oxford: Focal Press/Elsevier.

ITU-R. (1994). Multichannel stereophonic sound system with and without accompanying picture, Recommendation BS.775-1, International Telecommunication Union Radiocommunication Assembly, Geneva, Switzerland.

ITU-R. (1997). Methods for the subjective assessment of small impairments in audio systems including multichannel sound systems, Recommendation BS.1116-1, International Telecommunication Union Radiocommunication Assembly, Geneva, Switzerland.

Iwamiya, S., Nakajima, Y., Ueda, K., Kawahara, K., & Takada, M. (2003). Technical listening training: Improvement of sound sensitivity for acoustic engineers and sound designers. *Acoustical Science and Technology, 24*(1), 27–31.

Kassier, R., Brookes, T., & Rumsey, F. (2007). Training versus practice in spatial audio attribute evaluation tasks. *Proceedings of the 122nd Convention of the Audio Engineering Society*, Vienna, Austria.

Kidd, G., Mason, C. R., Rohtla, T. L., & Deliwala, P. S. (1998). Release from masking due to spatial separation of sources in the identification of nonspeech auditory patterns. *Journal of the Acoustical Society of America, 104*(1), 422–431.

Letowski, T. (1985). Development of technical listening skills: Timbre solfeggio. *Journal of the Audio Engineering Society, 33*, 240–244.

Levitin, D. J. (2006). *This is your brain on music: The science of a human obsession.* New York: Dutton/Penguin Group.

Martin, G. (2005). A new microphone technique for five-channel recording. *Proceedings of the 118th Convention of the Audio Engineering Society*, Preprint 6427, Barcelona, Spain.

Mason, R., Ford, N., Rumsey, F., & de Bruyn, B. (2000). Verbal and non-verbal elicitation techniques in the subjective assessment of spatial sound reproduction. *Proceedings of the 109th Convention of the Audio Engineering Society*, Preprint 5225, Los Angeles.

Massenburg, G. (1972). Parametric equalization. *Proceedings of the 42nd Convention of the Audio Engineering Society*, Los Angeles.

Massenburg, G. (accessed August 7, 2009). GML 8900 dynamic range controller series III user's reference, www.massenburg.com/cgi-bin/ml/8900ref.html.

Miskiewicz, A. (1992). Timbre solfege: A course in technical listening for sound engineers. *Journal of the Audio Engineering Society, 40,* 621–625.

Moore, B. C. J. (1997). *An introduction to the psychology of hearing* (4th ed.). San Diego, CA: Academic Press.

Moorefield, V. (2005). *The producer as composer: Shaping the sounds of popular music.* Cambridge, MA: MIT Press.

Neher, T., Brookes, T., & Rumsey, F. (2003). Unidimensional simulation of the spatial attribute "ensemble depth" for training purposes. Part 1: Pilot study into early reflection pattern characteristics. *Proceedings of the Audio Engineering Society 24th International Conference*, Banff, Canada.

Neher, T., Rumsey, F., & Brookes, T. (2002). Training of listeners for the evaluation of spatial sound reproduction. *Proceedings of the 112th Convention of the Audio Engineering Society*, Preprint 5584, Munich, Germany: AES.

Olive, S. (1994). A method for training listeners and selecting program material for listening tests. *Proceedings of the 97th Convention of the Audio Engineering Society*, Preprint 3893, San Francisco.

Olive, S. (2001). A new listener training software application. *Proceedings of the 110th Convention of the Audio Engineering Society*, Preprint 5384, Amsterdam, Netherlands.

Opolko, F. J., & Woszczyk, W. R. (1982). A combinative microphone technique using contact and air microphones. *Proceedings of the 72nd Convention of the Audio Engineering Society*, Anaheim, CA.

Quesnel, R. (1996). Timbral ear-trainer: Adaptive, interactive training of listening skills for evaluation of timbre. *Proceedings of the 100th Convention of the Audio Engineering Society*, Preprint 4241, Copenhagen, Denmark.

Quesnel, R. (2001). *A computer-assisted method for training and researching timbre memory and evaluation skills.* Ph.D. thesis. Montreal, Canada: McGill University.

Quesnel, R., & Woszczyk, W. R. (1994). A computer-aided system for timbral ear training. *Proceedings of the 96th Convention of the Audio Engineering Society*, Amsterdam, Netherlands.

Quesnel, R., Woszczyk, W., Corey, J., & Martin, G. (1999). A computer system for investigating and building synthetic auditory spaces, part 1. *Proceedings of the 107th Convention of the Audio Engineering Society*, Preprint 4992, New York.

Saberi, K., Dostal, L., Sadralodabai, T., Bull, V., & Perrott, D. R. (1991). Free-field release from masking. *Journal of the Acoustical Society of America, 90,* 1355–1370.

Schellenberg, E. G., Iverson, P., & McKinnon, M. C. (1999). Name that tune: Identifying popular recordings from brief excerpts. *Psychonomic Bulletin & Review, 6*(4), 641–646.

Schroeder, M. R. (1962). Natural sounding reverberation. *Journal of the Audio Engineering Society, 10*(3), 219–223.

Shinn-Cunningham, B. (2000). Learning reverberation: Considerations for spatial auditory displays. *Proceedings of the 2000 International Conference on Auditory Display*, Atlanta, GA.

Slawson, A. W. (1968). Vowel quality and musical timbre as functions of spectrum envelope and fundamental frequency. *Journal of the Acoustical Society of America, 43*(1), 87–101.

Smith, J. O. (accessed August 4, 2009). *Introduction to digital filters with audio applications*, http://ccrma.stanford.edu/~jos/filters. Online book.

Stone, H., & Sidel, J. L. (1993). *Sensory evaluation practices* (2nd ed.). San Diego, CA: Academic Press.

Usher, J. (2004). Visualizing auditory spatial imagery of multi-channel audio. *Proceedings of the 116th Convention of the Audio Engineering Society*, Berlin, Germany.

Usher, J., & Woszczyk, W. (2003). Design and testing of a graphical mapping tool for analyzing spatial audio scenes. *Proceedings of the Audio Engineering Society 24th International Conference*, Banff, Canada.

Verfaille, V., Zölzer, U., & Arfib, D. (2006). Adaptive digital audio effects (A-DAFx): A new class of sound transformations. *IEEE Transactions on Audio, Speech, and Language Processing, 4*(5), 1–15.

Woszczyk, W. R. (1993). Quality assessment of multichannel sound recordings. *Proceedings of the AES 12th International Conference on the Perception of Reproduced Sound* (pp. 197–218), Copenhagen, Denmark.

INDEX